The Almighty Machine

How Digitalization is Destroying Everything That Is Dear to Us

The Almighty Machine

How Digitalization is Destroying Everything That Is Dear to Us

Pekka Vahvanen

Translated by Mark Jones

zer0 books

Winchester, UK
Washington, USA

JOHN HUNT PUBLISHING

First published by Zero Books, 2021
Zero Books is an imprint of John Hunt Publishing Ltd., No. 3 East St., Alresford,
Hampshire SO24 9EE, UK
office@jhpbooks.com
www.johnhuntpublishing.com
www.zero-books.net

For distributor details and how to order please visit the 'Ordering' section on our website.

ISBN: 978 1 78904 898 8
978 1 78904 899 5 (ebook)
Library of Congress Control Number: 2021932920

A CIP catalogue record for this book is available from the British Library.

Design: Stuart Davies

UK: Printed and bound by CPI Group (UK) Ltd, Croydon, CR0 4YY
Printed in North America by CPI GPS partners

Contents

Preface

How can I believe that digitalization is destroying everything that is dear to us if I use a computer to write a book? Many have leveled that question after hearing the subtitle of this one. To those gnawing on the issue, I answer thus: even a soldier bogged down in the hell of battle might think that weapons destroy everything that is dear, but he nonetheless employs his machine gun in defense of himself and his land. Literary self-defense against the destructive force of computers is most effectively accomplished with the computer.

Modern technology provides power and dominion. Abstaining from it is therefore difficult when everyone else is using it. If I didn't exploit the word-processing function of the computer or gather information from the internet, penning a book would be slower and more laborious. Set against history, computers haven't made books any better. The book once held a far more substantial place in our culture. They brought more joy and thought to human life than they do now. Fifty years ago at least they were still read considerably more than they are today, despite perhaps not always being subjected to the meticulous rigor that computers bring to editing.

On a purely numerical level, technology can generate a competitive edge, but that doesn't automatically mean society or culture is better on the whole. Streamlining things rarely enlivens the experiential world of the individual or cultivates happiness. For the human, the past few centuries have seen the good parts of technological progress triumph over its bad; with technology we've pulled ourselves out of shortage, hunger, and the cold. The digital revolution, however, no longer improves the conditions in which we live.

This is why I have asserted that life overall would be better had the whole machine never been invented. My own life, though, would be worse had Mika Maliranta never been invented. I engaged him regularly in conversation on the necessity of progress, the development of artificial intelligence, and technological unemployment. I recall us on every occasion being in disagreement. Mika Kukkonen has also made the world a more enjoyable place to live, and he made this book somewhat better with his highly insightful comments on the manuscript. I want to thank Risto Heiskala as well for taking me in as his doctoral student in the summer of 2014. He provided invaluable conversational guidance for my writing both prior and after. Signs that the thesis is moving forward are still nowhere to be seen, but a bit of the mental work that went into it has found new light in this book.

I also want to thank several others who offered assistance over the course of its composition. They are Mona Moisala, Riikka Kämppi, Juha Nurmi, Minna Ruckenstein, Erka Koivunen, Juha-Matti Gärding, Sari Sarani, Matti Apunen, Juhana Vartiainen, Theodore Kaczynski, Sonja Saarikoski, Heikki Ervasti, Tanja Nuotio, Reijo Aarnio, Niina Junttila, Juho Toivola, Michael Laakasuo, Pekka Appelqvist, Torsten Groschup, Teppo Sillantaus, Hannu Hallamaa, Mari Kapanen, Maiju Lempinen, Kati Laukkanen, and many more.

Additionally, there will always be a special place in my heart and a debt of gratitude for those who stood behind the writing of a young guy in the midst of the uncertainty and angst swarming around the tasks of getting into university and finding a job. That guy, whose youth has already passed, hereby dedicates this book to Laura Saarikoski, Seppo Pänkäläinen, and Pekka Lakka.

It goes without saying that all errors possibly present in the book are the fault of machines.

Lappeenranta 31 August 2018

Pekka Vahvanen

1. Introduction

On Technology's Terms

But lo! Men have become the tools of their tools.
Henry David Thoreau[1]

One must go forward – step by step further into decadence.
Friedrich Nietzsche[2]

Technology is worshiped more intensely than any god. There is hardly holier writ in our time than the necessity of digitalization. Its believers are the billions infatuated with their smartphones and inextricable from the net. Its believers are politicians on both sides of the aisle who see technological advancement as the only way to keep their country alive.

All proclaim one liturgy: we must further Artificial Intelligence (AI) in order to ensure success among international competition. We must digitize primary education so that our children can study with modern methods. We must become a digital citizenry so that in the future we can stand on our own without anyone's help.

Our faith in technology is powerful because it has saved us in the past. From the Enlightenment onward, the world's most cherished ideologies have expounded the notion of social progress. Technology has remained a condition for the forward advancement we hold as inevitable, and something science-fiction author Arthur C. Clark called "indistinguishable from magic."[3] The Industrial Revolution took off at the end of the eighteenth century and launched humanity from its place at the subsistence level into unseen prosperity. The late nineteenth and twentieth centuries made the magic of technology part of our daily existence – everyone now had the sorcery to travel from place to place rapidly and talk with people on the other side of the planet.

Our belief in the benevolence of technology, however, is beginning to flirt with gimbal lock. Technology is no longer a force that molds society into something better, and this book is an examination of how advancement has managed to depart from its intended course. We are endowed already with too much progress, and it no longer contributes to our well-being. The things that once powered us toward a brighter tomorrow are now undermining our purpose.

A quiet opposition to the satiating nature of technology has begun to take shape. Intel's research division, IntelLabs, conducted a survey that showed 61 percent of young adults feel technology is making the world less humane. The responses from eight nations indicate how more than anyone, the millennial is troubled by the undesirable sides of tech.[4]

We have indulged in the joys of social media, but we have also become drenched in its deleterious effects. The aversion to meeting and being with each other just grows as internet messaging makes communication "easier." The subtle nuances of face-to-face interaction and our capacity for empathy get crowded out in internet conversation. Social media is a wellspring of fabricated happiness where the constant comparison with others' performance and appearance breeds feelings of insufficiency. Research links time spent on Facebook with unhappiness, and it shows that just setting it aside allows satisfaction with life to increase.[5]

The disappearance of jobs is also a point of concern. The oft cited work of Oxford University researchers Carl Benedikt Frey and Michael A. Osborne predicts that AI and robotics development could eliminate 47 percent of jobs in America by the early 2030s.[6]

The 2016 US presidential election revealed how the digital transformation is also having its way with politics. Political consulting firm Cambridge Analytica used farmed data to construct voter models that made manipulating voters easier. Many advertisers do the same – stick their noses in our search terms looking for strings to pull to get us to buy more trinkets. Facebook has been less than eager to regulate the violation of privacy or the manipulation of users – the very pillars on which its business model is based.

Our ability to concentrate and engage in reflective thought is also under threat in the computer age. It is not uncommon for office workers today to check their email and click through Facebook pages tens of times every hour.[7] This makes the execution of tasks that require prolonged thought more difficult, and in the face of the flood of stimulant, new and immediate are often prioritized over important.

Intelligence researchers are already asking if the flattening (or in Scandinavia, the decreasing) IQ test scores of the developed world are partly due to smartphone and computer culture. Nobel prize-winning economist Robert Solow has famously stated that the rise of computers has been a boon for everything – except productivity. That figure for Finland from 2007-2017 has remained essentially unchanged despite constant growth in computer capability. Could this gap be at least somewhat attributable to computers making *us* lazier, stupider, and less industrious?[8]

In spite of it all, the economic and political elite see no other alternative than leaping headlong into the stream of digitalization.

And many have raised red flags. More extreme than the threat of mass unemployment are the long-term scenarios where an AI that exceeds human intelligence destroys civilization. Fewer and fewer thinking humans relegate this to the realm of science fiction; it has been accepted as a genuine danger by such names as physicist Stephen Hawking, Tesla CEO Elon Musk, and Oxford University AI expert Nick Bostrom.[9]

Some of technology's effects are unequivocally at odds with us. The idea that our digital habits are spied on and that our data is being used against us will never sit nice. Technology is seductive, and we use it completely on our own volition. It is easy to become addicted to Facebook, because it whets our appetite for admirers and self-emphasis. Simultaneously, we wake up to the realization that it is toxic.

Two well-known dystopias that center on technology emerged in Britain before the world was halfway through the twentieth century. George Orwell's *1984* has seen more of the limelight. In Orwellian society, nearly every movement of the individual is monitored, freedom is crushed, and a suffocating state enterprise employs technology to control and subjugate its citizens.

Aldus Huxley's nightmare is subtler, but in no way less worrying. The people of *Brave New World* use efficiency and enjoyment as their benchmark for everything, all things seem to be in place, but closer inspection uncovers a twisted picture of progress and happiness. Humanity's needs are met and life appears as one uniform plateau of pleasure, but people live decoupled from nature, and (almost) everyone persists in a state of domestication, incapable of thinking outside the lines or imagining anything bigger than themselves.

If Orwell feared the censorship of books, Huxley feared that no one would even care to read. If Orwell feared some external treachery would make life a living hell, Huxley feared a life of abundance and fun would make our lives meaningless.[10]

Huxley's nightmares are increasingly our own. Much of the substance of life goes away when we approach it through the lens of the IT engineer. We are prone to maximizing the principle of efficiency, scraping ourselves spiritually hollow, chasing after empty entertainment, and becoming those for whom a hexadecimal world of no diversity is unenjoyable.

/ / /

There is an ugly side to progress that we are not invited to see, because many have an entrenched interest in keeping the perpetual development engine humming at high rpm.

But there is much to behold on this ugly side. Whenever technological advancement gives us something new, it takes something else away. Industry and transport systems that are dependent on fossil fuels have brought us wealth and well-being – i.e., progress, but this forward march has polluted the atmosphere, contaminated the waterways, and (as a matter of general scientific consensus) is jeopardizing the future of our whole planet.

Progress in the development of IT is also backfiring on a number of levels. Devices that made knowledge more accessible than ever are now making us more ignorant. Machines that enhance efficiency are turning us into bumbling idiots – they promise more economic well-being while diminishing it for many, and while bringing us more connectivity than ever they are poisoning the connections between us. The bytes that profess to rid life of its monotony are precisely the ones that are robbing it of meaning, and the contraptions meant to expand our capacity to harness our surroundings are training us as slaves – and could destroy us on the way to taking earth's throne.

The Machine and Me

All of us come uniquely disposed with a set of sympathies and antipathies from our own life experience. It is my belief that writers and researchers have a duty to convey these to the reader. Those claiming to speak from a position of pure objectivity are lying.

I want to make it clear that I have never had a particularly affectionate relationship with computers. That goes for social media too, and it is also worthy of mention that my phone is an old Nokia with no access to the net. I am a fossil of an earlier age – a Luddite, many might say.

As we are not foundationally rational beings, I have tried to search my life history for an explanation behind my negative outlook on technology. A few things have turned up – issues that cast a shadow on my connection to tech (and certainly computers) from the very beginning. When I was a child, our home computer was a tool for playing games. My father rambled on at length about how computers were going to change bookkeeping, writing, and blah blah blah. None of this mattered to me, and I was neither consumed with a desire to play computer games, although I had some fun with them. There was just one problem – in order to get the game to run, I had to sit with my father in front of the box for hours. This was boring. I had no idea what he was doing, and staring at the display quickly turned my brain into porridge. When it finally got going, the neighborhood boys wanted to come over and play it with me. I did not like this, because I did not like the neighborhood boys.

Our first computer was a PC from the early 1990s called the Tandy 1000. Next, we had a relatively more advanced IBM 386 MHz machine with 2 Megabytes of RAM. Later, we purchased two more Megs to get our F1GP racing game to play. Today's PCs are typically fit with thousands of times more memory than that IBM.

Computers are designed to make life easier, and sometimes they pull that off, but not always. My current PC often restarts itself in the middle of something in order to install necessary updates, which it then fails to do. I am forced to wage perpetual war with its word-processing program because it thinks it knows better than me what term I want to use. One would think that at least at the gym one might find release from the digital despotism that rules our lives, but already during warm-up, we must wield some degree of computer engineering expertise to find the submenu that starts the damn exercise bike.

A stockpile of pent-up frustration would not have been the sole source of motivation for writing this book. My more rational worldview took a jarring after receiving my Master's degree in history in 2011 when I watched a documentary that gave me my first glimpse of Ray Kurzweil. This current Googler and prize-winning inventor was soliloquying expressionlessly on my screen about a future where we merge with machines. Kurzweil recounted about how AI will exceed human intelligence by 2045 – after which who knows where things will go. In Kurzweil's opinion, super intelligent machines will only make the world a better place – accelerating scientific and economic progress to levels heretofore unimaginable and conquering disease with whatever these overwhelmingly superior beings create.

What disturbed me was the other possible course of events that Kurzweil illustrated – the one where machines with extra-human intelligence end us. Kurzweil is trying to get humanity to take the most dangerous gamble in all of its history. The question is, do we want to double down?

In 2011, my review of Ian Morris's *Why the West Rules – For Now* ran in *Finnish Foreign Policy Magazine*.[11] While searching through the book for an historical analysis of the economic and political rise of China, I came across some surprising conclusions in its final chapter. Morris was asserting that the power play between East and West – i.e., China and the US – could fade away into irrelevancy in the future due to machines smarter than us diluting the whole purpose of power struggle.

My coursework in college ignited an interest in international affairs, and I believed at the time that the world's superpowers and their leaders are what ultimately determine the fate of people. For my Master's thesis, I enthusiastically engrossed myself in the strategy of Richard Nixon's foreign policy, but by 2011 I had begun to see that the tug of war between states – if not totally irrelevant – would soon be more akin to a sideshow.

While this view is not shared by all, it is by no means a new idea. Many historians in recent decades have criticized history written only about prominent statesmen, and some have highlighted the pivotal influence of technology. English writer Joseph Glanvill wrote in the 1600s that we owe more thanks to the unnamed inventor of the marine compass "than to a thousand Alexanders and Caesars, or to ten times the number of Aristotles."[12] The tidal wave of technology has swollen in size since the first years of the Industrial Revolution, and today, instead of politicians, priests, and generals, our lives are more under the thumb of engineers, inventors, scientists, and CEOs.

The profundity of the change undergone by the human community in the last 2 centuries is evident from Morris's graph of social development. Until 1800, little disruption to human life is visible when viewed through Morris's analysis of energy capture, urbanization, IT, and war-making capacity.[13] As a consequence of the Industrial Revolution, change came quickly after 1800, mainly due to unprecedented strides in technology. Engineers changed the world.

The roles played by statesmen and clerics in social

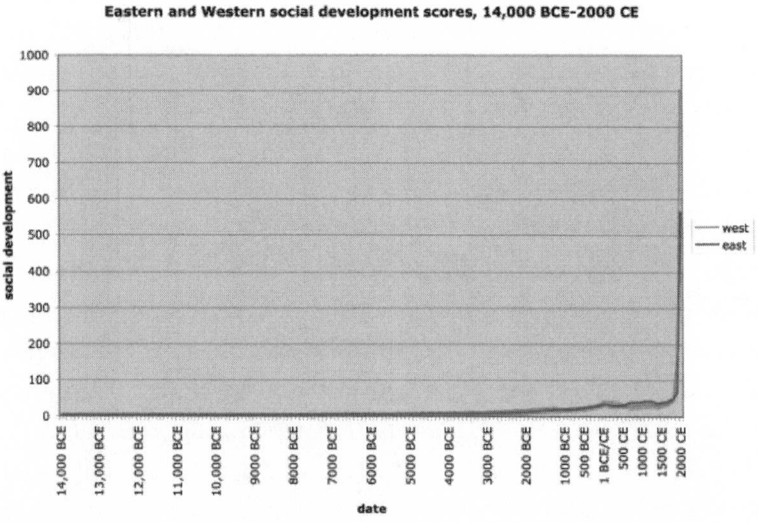

development were not insignificant. Tunneled religious views and debilitating wars served to stifle the momentum of development. (The idea that the buildup of the war machine propels the general advance of technology downwind of it was, at least prior to the twentieth century, not plausible.)[14]

Of note, however, is the fact that the tyrants of the 1900s and 2000s were rabid fans of technology. All nations today, the isolated North Korean dictatorship included, are trying to maximize their might by drafting leading-edge tech into state service. Although Donald Trump came to power partly on the vote of those who lost jobs to technological advancement, he is not against it. Previously one of his staunchest supporters and advisors from the Silicon Valley brilliant boys club, billionaire Peter Thiel, adamantly insists that with technology we can deliver man from all of his problems – even death.[15]

While technology enjoys immense influence over us, governments have limited means with which to control its progression. New gadgets just get embedded in our everyday without anyone caring to question them. Futurist Alvin Toffler submitted in the 1970s that industrial societies need a governing body to debate the potential social effects of new technologies before they make it to market. Toffler's take was that, "We must stop being afraid to exert systematic social control over technology."[16]

It is customary for anyone pushing change to bear the burden of proof for why it is essential, but in tech development things apparently work the other way. Those that are behind the devices disrupting our lives are always right, while those supporting the status quo are charged with presenting convincing arguments against them. Law rarely constrains the engineer.

The Weight of the World on His Shoulders

I come from the Ironic Generation. Reserving nerves for social issues in a world that predates us might sound instinctively silly, but I freely admit that technology's ability to obliterate precious pieces of life has been worrying me.

In 2012, the writing I did for newspapers and periodicals began to focus more on how new technologies are swinging society's rudder – often in a more *unsettling* direction. These include articles addressing IT's destruction of thought, the potential for technological unemployment, killer robots, and machines more intelligent than humans. The first to be run under the bus of my viewpoint were the readers of the *Sunnuntaisuomalainen* and *Suomen Kuvalehti*, and later the *Helsingin Sanomat*. I have also spread my antitechnologist heresy on the prime-time Finnish talk show *Perjantai*, of which I was a host.

The talk around town has undergone a shift, and the detrimental influence of new tech is now being given more attention. In 2013, a story of mine about technological unemployment was scoffed at as too speculative by those to whom I initially submitted it. For the story, I had interviewed Martin Ford (mentioned later), an author who believes technology in the coming years will cause mass unemployment, as well as Finnish Labor Minister Lauri Ihalainen, who had little regard for Ford's opinion. Today, press about the threat of technological unemployment graces the pages of the media basically every month.

In 2013, I also exchanged letters with the notorious anti-tech terrorist Theodore Kaczynski, also known as the Unabomber. A piece summarizing our conversation appeared in the March 2015 supplement of the *Helsingin Sanomat*, the most widely read newspaper in Finland. I was especially intrigued by Kaczynski's often razor sharp – although somewhat extreme – critique of technology, a subject on which he has written two books while in prison. In no way have I ever tried to eulogize the acts of carnage that he committed, which over the course of 17 years took the lives of three people and injured more than 20.

The psychiatric evaluation Sally Johnson performed on

Kaczynski in connection with his trial is interesting from the standpoint of this book. Her pronouncement of Kaczynski's opposition to technology as being in itself proof of his mental incapacity reveals much about modern society's belief in progress.[17] Today, as people walk the streets fixated on their smartphones, and while digital devices occupy our lives from sunup to sundown, opposition to technology is already far more justifiable than 2 decades prior when Johnson wrote her report.

The central themes of this book just a decade ago would have seemed less credible to many readers. Today, however, research reveals quite indisputably how continuous smart device and social media use are corroding our cognition. A conspicuous contingent of mainstream economists is now certain that digitalization is fostering inequality and threatening to bring about mass unemployment. Sociologists have awoken to the radical way in which virtual life alters our relationship to other people, as well as the capacity of new tech to fuel the flame of mental illness. Nearly every person exposed to the news has seen how the collection of big data through privacy violations enables a more refined form of manipulation, and the most influential figures of the modern world are talking about how playing with the AI fire might leave us burned.

Dissertation

This book owes its origin to the dissertation work that I began in spring 2014, the idea being to dig into what members of the Finnish parliament think about the social consequences of AI. Several of my professors were of the opinion that there is not much to uncover there, as legislators know next to nothing about the subject. I have never questioned the infallibility of the beings that lead us, but my dissertation has nevertheless gone nowhere beyond outline, namely because hosting TV and radio shows has taken up most of my time. I also decided that if I were to write a non-academic book about the matter, there might actually be people that want to read it. It was my belief that something aimed at the general public would let me say much more than I could ever get away with in a thesis. And that is what I have done.

At the centroid of this book sits a question of progress: does technological development also bring social and moral steps forward? Does it improve human life, and does it help us in the pursuit of happiness?

The idea of progress – its history and future – is addressed more deeply in Chapter 2, where I introduce a concept central to the rest of the book – the *progress trap* – i.e., how exorbitant social (like individual) achievement can turn on itself. I ask – are we becoming the casualties of our own success?

In the chapters that follow, I present five big issues (progress traps) that excessive advancement in IT and AI have caused, or are causing.

Chapter 3 examines how IT dulls our thought processes and our intellectual culture. We are losing hold on the heritage we built with the written word, whose virtues include a sustained focus and a capacity for reflective, critical thinking. Studies have shown that we care to read from the internet far less than we do from paper text in one sitting, and there is evidence that suggests our ability to learn in a multimedia environment is inferior to a literary one – where we need not continuously click on things. Nicholas Carr has said of the internet's design that it is above all a place of interruption and interference. It is one that could make us more superficial and polarized, and our society consequently less democratic.[18]

In Chapter 4, I discuss the demise of privacy and the potential for manipulation that this enables. Edward Snowden, the well-known whistle blower for the spying programs of the National Security Agency (NSA), has said the surveillance of Orwell's *1984* pales in comparison to what really takes place. Many wonder why you would even worry if you are not doing anything illegal. The reason: our data is being passed around in a long list of shady ways. The situation is the same as the prototypical nabbing of the bad guy in the classic American movie: everything you say can and will be used against you.

Chapter 5 sets out to search for purpose – that which our lives no longer contain. The digital devices that are taking a growing portion of our time water down what is meaningful and monocularize the experience of being human. We are on the verge of devolving into people who no longer dream about, shoot for, or embark on anything. Sociologist Sherry Turkle has characterized the influence of social media by saying we are both alone together, and together alone. We are no more capable of being locked up with our own thoughts today than we are capable of being at ease with other people. Younger age groups in particular exhibit a growing lack of interest in the notion of face-to-face interaction, and the smartphone generation has been shown to be coping more with psychiatric instability.

In Chapter 6 we look at the threat of technological unemployment. The flip side to advancements in IT and AI is that in the long run more of us will be without a place in the job market. The attitude and effort of the individual is rendered irrelevant if in the future machines can do almost everything more effectively. A broadening legion of people are becoming worthless to both the economy and society.

We are often treated to the line about how the looming doom of technological unemployment has been with us forever – a point which is true, as it was wandering around already before the early-nineteenth century Luddites. Today, however, things are completely different – AI is filling requisitions for the most highly educated professions. New roles and positions will undoubtedly emerge, but seemingly nowhere close to the amount needed to account for those that are vanishing. Burgeoning firms in the IT field employ less than old industry. This chapter also discusses how even before any potential mass unemployment, the rift between the endowed and the impoverished will widen. The digital economy is borne on a vortex of inequality. "To all who have, more will be given, and they will have an abundance; but from those who have nothing, even what they have will be taken away."

Chapter 7 examines the catastrophic potential of new technologies. Ever since missiles began to be fitted with hydrogen warheads, we have had the capacity to nearly rid the world of the human race. The military and economic brawl between world powers encourages us to amass large inventories of small planetary disasters. Many nations today are building AI-guided autonomous weapons that are given kinetic employment authority in life and death decisions. AI's arrival at the intelligence level of people could be our final invention – the last thing we ever do.

In Chapter 8, we ask if there are any alternatives to voluntarily digging our own grave with digitalization. Debate about tech development has found its way into the politisphere before, but the defining ideological dilemma of this century will likely be what we do to limit science and technology. This issue will shortly garnish more political visibility, because friction between the tech elite and those displaced by their creations will grow more volatile.

Without question, there is more to the dark side of digitalization than that presented in this book. Net crime and online bullying, for instance, are already rampant problems without resolution. They are not new – they just moved onto the net. This book also leaves unaddressed pressing issues of cyber security, despite the immeasurable distress and billions in expenses that they create. Instead, I examine the all-pervasive human conundrum that we face through the eyes of the big five life-disrupting shifts in our existence previously stated.

Both stricter and more liberal interpretations of the terms *digitalization, digitization,* and *digital transformation* are adrift in the discourse. This book uses the word digitalization to mean the conversion of various areas of life into a format that is digital, including those that are social, intellectual, economic, political, and military. Digitalization colonizes every corner of our lives, everything that we know is migrated onto the web, and the influence of this digital truth extends into the world beyond the bytes.

The reader can pick and choose chapters as if each was an independent essay formulated to stand on its own. The book's contents, however, will of course have the most efficacy when read in full, because only then will the reader get a full taste of the senseless dead-end that is digitalization.

/ / /

We humanists often lack the confidence to trudge into the subject of technology. Our fear is a lack of knowledge, and also that technical matters to many of us occupy the territory of tedium. Those that elect to speak and write about tech are therefore those that are intimately familiar with and proponents of it. The danger here is that technology's beveled edge is presented in a way that conceals all but its positives. The late cultural critic Neil Postman once said that one need not be an auto mechanic to see how the automobile has changed urban planning. Similarly, one need not be a coder or IT professor to see how the digitalization of everything is changing how we think, what society is, and what it means to be human.

I have tried in this book to give technology credit where credit is due, but as is no doubt already clear, this book is intended to serve as a counterweight to digi-enthusiasm. The only fix for a ship that is badly listing is to add ballast to the other side.

So, first let's make a run through history, in order to comprehend what we are up against today. In the next chapter, we take a look at how technology has brought improvements to human life in the past. This force that once made our lives better has in recent years managed to run itself dry, and in the future, its underlying risks and issues will likely outweigh its benefits.

2. Progress Traps

Technology once made our lives better, but its limits are now upon us

Our entire much-praised technological progress, and civilization generally, is comparable to an axe in the hand of a pathological criminal.
Albert Einstein[19]

[T]he future, which has so many elements of high promise, is yet only a stone's throw from despair.
Robert Oppenheimer[20]

The anticipation of a better tomorrow – the dream of what looms on the horizon – is often what infuses life with purpose. Children wait with impatience for the day when they can finally be adults, because then they can pick their own holidays and drive a car. The bliss of youth is built on the allure of future promise, the fulfillment of which one need not bother to wait for. We remain capable still in old age of greeting the arrival of retirement as the gateway to our golden years when we finally have time to focus on the things that we find important.

The modern human enlists *progress* as the rubric for his life. We envisage ourselves as sailing linearly on into sweeter seas – if not in our private lives, then at least in our professional. Research into individual happiness shows that people typically consider their present and future to be better than their past. The same research, however, indicates that the happiness experienced by an individual doesn't exactly change much over their lifespan.[21] Meaning, we paint castles in the sky about always being on the cusp of another nirvana but eventually discover we were just as happy before.

Equally widespread has been the belief that the trajectory of society is perpetual ascent. The notion, though, that societies can be made richer and happier through science, technology, and politics is relatively new, having only gained traction in the eighteenth century among the elites.

Throughout classical antiquity, it was frequently believed that the golden age had come and gone, and people just braced for more poignant forms of crucible. Change was met with skepticism. It was also commonly held that history unfolds in cycles – that everything keeps repeating itself without fundamental change. This cyclical view helped disincentivize believing that society will always improve.[22]

Multiple early epics and accounts in the Old Testament warn against dipping one's fingers too far into progress. They speak of the vengeance that awaits those whose curiosity, thirst for knowledge, and proclivity to play God balloon beyond excessive. Prometheus gave man fire, but Zeus gave him something else to punish this technological transgression: woman. The first woman, Pandora, was drunk with curiosity and a thirst for knowledge, and proved unable to resist opening the box and setting all the world's evils free. As historian J.B. Bury writes, the ancient Greeks knew that "subtle dangers lurked in human achievements and gains."[23]

Adam and Eve's adventures in the story of creation got them ejected from paradise for partaking of the fruit of the tree of the knowledge of good and evil against God's decree. The Tower of Babel was constructed to reach heaven, but God reduced it to ruins and returned those clamoring upon it to earth.

These fables foster the thought that the accumulation of knowledge, the development of technology, and the altering of the environment run counter to God's will. But the biblical story of creation enables another interpretation – one where man is endowed with the right and duty to "cultivate and care" for the land – to harness nature at God's command.

Historian Joel Mokyr writes that this condonement for the refashioning of the environment gradually began to walk on its own in the Middle Ages. The Benedictine monks thrust it into full gear by extolling the virtuousness of hard work. God soon took on the stature of architect or engineer, which explains much about the shift in values during that time. So, just when those far more sophisticated than Europe in the Islamic world began to approach technology with reservation, a Christian Europe welcomed its development with open arms.[24] The cherubs guarding the gates to paradise be damned. Onward to the mythic fruit trees of Eden!

Advancing Toward Perfection

When we talk about progress, we are usually referring to one of three things that our modern perspective intertwines into one hybrid. The first is scientific and technological progress, which enhances man's prowess at taming nature. The second is society's material well-being, which is largely descendant from the achievements of science and technology. The third is the grand moral progress tied to the sociopolitical crusade for freedom and happiness.[25]

Many today take it as self-evident that science and technology can help improve our material standard of living. Instrumental in the formulation of this idea was the modestly titled "genius" in encyclopedias known as Francis Bacon (1561-1626). According to him, science was never meant to serve as a stack of speculation stemming from a thirst for mere knowledge; its primary motivation ought to be the reformatting of nature into congruence with man's prerogative.[26] Bacon's rationale was shared by many philosophers of the Enlightenment.

The most profound thinkers of the 1700s no longer viewed history as some blundering stagger. It was now a ticker-tape parade into the blinding brilliance of tomorrow. Knowledge, science, technology, liberalist principles, and good governance were the hidden hands that set societies in line. The French economist and statesman Anne Robert Jacques Turgot (1727-1781) was one of the more vivacious wavers of the flag of progress. His belief was that despite its wars and its ugly catastrophes, history persists as a fairly unwavering climb into illuminance. Progress to him meant the transition from hunter-gatherer into pastoral herder, then agriculturalist, and that later, perspectives were further widened by exploration and commerce. Turgot submitted that a migration had also begun to appear within government – away from despotism to a more humane form of statecraft and broader freedom.[27]

Even the historian, skeptic, and author of a revered work detailing the wither and collapse of Rome, Edward Gibbon (1737-1794), became infected with the mantra of his contemporaries and began spreading the progress gospel. He bade his readership "acquiesce in the pleasing conclusion that every age of the world has increased, and still increases, the real wealth, the happiness, the knowledge, and perhaps the virtue, of the human race." The enchanted historian adds that, "[W]e cannot determine to what height the human species may aspire in their advances toward perfection; but it may safely be presumed, that no people, unless the face of nature is changed, will relapse into their original barbarism."[28]

A powerful belief in progress was facilitated in part by the fact that significant developments had occurred in preceding centuries. These show us that man's ability to leverage nature has undergone growth. While the Middle Ages are not primarily admired for being a fireworks display of innovation, they gave birth to the printing press, the mechanical clock, gunpowder, and the European version of the compass. (The Chinese had every one of these already up and running in some form before Europe.) The utilization of hydropower expanded in bounds, windmills were invented, and metals could now be extracted with ease. Open ocean sailings were made possible by more dependable shipbuilding and navigation techniques.[29] These forward strides lent themselves to the belief that we were living in more advanced times than the ancient Greek philosophers.

The prevailing ethnocentrism of the age aside, some scholars proved progressive enough to include the people of other cultures in their ever-widening classification of progress. For example, the French encyclopedists believed that all races and nationalities can refine their societies nearly endlessly.[30] Indeed up to and even beyond the Second World War, non-European peoples were patronizingly coaxed onto the road of development.

Despite this ardor, a hard truth must be stated: material progress was not that bad before the Industrial Revolution. Hunter-gatherers fed themselves well in comparison to the richest societies of the early modern world. The poor of England in 1790 often subsisted on 1500 calories per day, which is less than our forager ancestors. The general life expectancy before the Industrial Revolution in England was shorter than that of our forefathers who clubbed their prey, and who after making it through childhood are known to have lived well into their sixties and seventies. The Industrial Revolution brought about a dramatic rise in these figures. After 1850, the average life expectancy grew globally by nearly 3 months per year, and since 1800, the world's wealthy economies have become 10-20 times richer.[31]

It was technology – namely those led by the steam engine and the spinning jenny – that put power behind the Industrial Revolution. The presence of new equipment yielded newfound economic efficiency. The annual productivity increase during the centuries before the Industrial Revolution was not even 0.05 percent, but at the end of the eighteenth century it was already growing at half a percent per year – more than ten times what it was earlier. Despite this, the quality of life for workers remained dismally unchanged until the middle of the nineteenth century.[32]

The Industrial Revolution, or "the machinery question," brought long workdays and a cramped life in a polluted city. The Scottish essayist Thomas Carlyle highlighted how after the arrival of machines, society began to function almost as if it was one: it became reminiscent of the very herald of change, the steam engine "rolling on in its dead indifference." Society saw also a similar rise in regard for only those facets of life that could be quantified, calculated, and manipulated.[33]

Opposition to the emergence of machines in many parts of England had grown so potent that some skilled laborers elected to destroy their power looms themselves. "[T]hese sufferers hated the machines which they believed took their bread from them, they hated the buildings which contained these machines, they hated the manufacturers who owned those buildings," wrote Charlotte Brontë. The deeds of these so-called Luddites were often denounced as senseless and futile, given the long-term benefits automation gave to society, but that brand of foresight did not have a place in the priorities of the laborer. Historian Eric Hobsbawm alleges that the Luddites in fact achieved what they set out to do. Their method of "collective bargaining by riot" delayed the introduction of automated machinery in many locations, and in so doing fortified the worker's standing in the market. For them, it was a victory of meaning.[34]

Any aggregate assessment of the Industrial Revolution will reveal it to have fueled one of the most important course changes in human history: the unshackling of ourselves from the Malthusian Trap. This famous theory propounded by Englishman Thomas Malthus postulates that a slight rise in living conditions will generate an increase in population, while overpopulation will eat up the resources required to set the standard of living higher (thereby often leading to famine). In other words, what goes up must come down. But things haven't quite transpired as forecast. Our population has of course grown, but our standard of living has risen with it. Mechanisms limiting economic growth appear to have vanished, along with other constraints purported to be inherent. The railroad shattered our perceptions of time and place during the latter half of the nineteenth century. Better food, hygiene, and more advanced medicine improved our health and blazed the trail for everyone to be elderly. Like Herbert Spencer, more and more people were convinced that rather than a fluke, progress was a necessity.[35]

Although the relationship between man and the environment underwent significant metamorphosis in the nineteenth century, the material development in the one that followed altered the life of the layman even more drastically. The period spanning the years from 1950-1970 that we call the golden age saw a myriad of conveniences – from the automobile and telephone to the refrigerator and television – become permanent fixtures in our version of normal. The influence of the washing machine was revolutionary: it liberated women from homemaking and paved them a lane into the job market. In the 1980s and 1990s our homes began freeing up space for computers, which until then had found use only within the defense sector, universities, and corporations.

The typical worker of today in many ways enjoys a more lavishly abundant lifestyle than kings of previous generations (although in developing nations absolute poverty persists, but continues to decrease). The baby-boomers rode one exceptionally rad wave of technological change.

Mothers of Necessities

It is often said that necessity is the mother of invention. When a need arises in society to invent something, scientists and innovators train their focus on whatever that is, and eventually things find their way into a solution as if wisely designed that way from afar. In some cases, things generally unfold in this fashion – James Watt's steam engine solved the water pumping problem in English mining.[36] But the process of devising new technology does not always come replete with a clear picture of what one is devising, what can be done with it, and how necessary it is to everyone. It is very difficult to predict (and impossible to control) the long-term effects of any innovation.

Sometimes we might set out to find India, but then end up discovering America. The potential for the use of X-rays in the medical imaging of patients was uncovered by accident (1895) while Dr Röntgen was conducting experiments with Crookes tubes. The transistor came about (1947) somewhat unintentionally as a byproduct of research into semiconductors. Many things that we consider essential today often previously appeared worthless in the eyes of scientists. When chemists in the 1800s derived fuel from crude oil that was usable in lamps, they pronounced this benzine byproduct to be nothing but waste – then everyone suddenly needed it.[37]

Few in the 1880s found it noteworthy when Karl Benz built the first car. Humans had for 6000 years relied on horses as their mode of transport. Steam locomotives, powered by the very steam engine created for coal mining, proved a splendid means for trips of long duration. Over time, the automobile ceased to be shunned as nonessential. Armies began to haul gear with trucks in notable numbers beginning in the First World War. The costs of these once prohibitively expensive and frivolous contraptions slowly came down, and soon the American, the European, the Asian, and the African had to have them.[38] The long-term necessity of the automobile was accentuated by the fact that getting by without them became difficult simply because the design of everything else was predicated on everyone having one.

Just when a new gadget is devised and successfully peddled as impossible to pass up, people start coming around to the conclusion that it is essential. Rarely is necessity the mother of invention – invention is rather the mother of necessity, as geographer Jared Diamond puts it.[39] Far less often than we might think did people in the nineteenth and twentieth centuries find it necessary to photograph their lunch, but today's mobile phone cameras have apparently made this a requirement. Before electricity, people probably didn't sit around holding their breath for science to solve their "problem." Today civilization shuts down without power.

Whatever our inventions eventually turn into will surprise everyone as long as our world remains chaotic. Neither Karl Benz nor Henry Ford could have foretold the immensity of the political upheaval that would come from the oil hunt. They could not have imagined that automobile accidents would take more than a million lives annually, but also save many; young and healthy traffic casualties lost to head injuries are the ideal organ donors, and it's thanks to the ambulance that accident victims and others suffering complications can be gotten to care with speed.

Today the progression of science and technology comes off as a lip-locked promenade of lovers, but the ornate sheen of new science was not always the bedfellow of invention. In the Middle Ages, new technology usually came about through trial and error. As mathematician and philosopher Alfred North Whitehead states, Europe still knew less about science in 1500 than Archimedes did in 212 BC. Simultaneously, Europe was technologically far ahead of anything the ancients had achieved.[40]

The most significant inventions of the Industrial Revolution are only marginally attributable to science. The scientific method led James Watt and his colleagues to many discoveries, but the latest scientific knowledge did not play a leading role in their tinkering. Science proved to be a major influence in electricity-related inventions at the end of the nineteenth century, but it was in the next when the alliance between science and technology became undeniable.[41]

When the tip of the scientific spear began to take the wheel and drive development, technology fell under the purview of an oligarchy of experts. The task load of the scientist consisted of things no one in politics or the populace could make heads or tails of. The ethics expected out of scientists therefore grew, and the small cadre of physicists pursuing the prospect of nuclear energy could have acted differently. Had they refrained in 1937 from revealing the potential of nuclear fission to political and military leadership, the world would have remained blissfully unaware of nuclear weaponry – at least then.[42]

Technologies derived from cutting edge science became a growing component of the jousts between world powers – as evidenced by the proliferation of nuclear stockpiles and space programs. Government funding gobbled up greater portions of scientific research, because science had become the feeding tube for the state's existence. As futurist Alvin Toffler wrote, capitalists and socialists were united behind their unflinching belief in technology, the subjugation of nature, and the economic growth bubbling up from within.[43] The socialist bastion would perhaps have emerged unscathed in every way had it not failed to outcompete in technology and economics. The West was able to equip its citizens with a higher standard of living because of its overwhelmingly superior techno-economic engine – which, likely more than anything, brought the curtain down on the Cold War.

Over the last quarter century, computers have stolen the show of technological development. With the exception of a handful of prescient scientific minds, few foresaw their potential when Alan Turing published his first article alluding to them in 1936. The average among us had always gotten along fabulously until the turn of the twenty-first century without computers. Today most of us would call life without them an impossibility. If one wishes to survive now, it is imperative that one acquires fluency in computers.

Technological development in other disciplines over recent decades has been modest. Tech investor Peter Thiel, for instance, has bemoaned the lack of advancement in areas other than IT. He raises the case where developmental stagnation manifests itself as a drag on how fast we travel. The speed of sailing vessels increased continuously from the sixteenth into the eighteenth centuries. Train velocities went up throughout the nineteenth, and in the twentieth cars and aircraft just got quicker and quicker. But when the supersonic Concorde ceased operations in 2003, progress regressed into negative. In the 1970s one could fly from London to New York in 3.5 hours – roughly half the time required today. Thiel also famously lamented that "we wanted flying cars, instead we got 140 characters."[44]

From the standpoint of our day to day, it is difficult to imagine that we could ever drum up a more useful spectrum of inventions than those of the last two hundred years. Finnish novelist Hannu Raittila wrote that the flushable toilet can be invented only once, and that no correspondingly life-enhancing innovations are on their way. Britain's *The Economist* later ran an issue echoing the sentiment, with a cover of a Rodin Thinker sitting on the loo and a caption that asks, "Will we ever invent anything this useful again?"[45]

Man is Morally Better Today Than in the Middle Ages...Kind Of

It is clear that our ability to stand as kapellmeister over nature has powerfully enhanced our physical well-being. But has this material progress also brought about a moral one? As history bulldozes on, has it made us better people? Let's ask the philosophers.

"Nothing is more gentle than man in his primitive state," said the esteemed Genevan, Jean-Jacques Rousseau (1712-1778). Englishman Thomas Hobbes (1588-1679) rather found no reason to yearn for prehistory: "[N]o arts; no letters; no society; and which is worst of all, constant fear, and danger of violent death; and the life of man, solitary, poor, nasty, brutish, and short." To Hobbes, our pupation from caveman to contemporary was a massive moral progression, while Rousseau was convinced that "our minds have been corrupted in proportion as the arts and sciences have improved."[46]

History, though, does not go so far as to unequivocally crown either Rousseau or Hobbes as right. Perhaps we haven't yet shapeshifted into meaningfully better beings, or perhaps that isn't even a prerequisite for moral progress. The political and economic institutions of the modern world have provided incentives to prevent condemnable deeds, and dredged a canal connecting our greed with the greater good.

Selfishness, ambition, and vainglory, wrote Turgot, are the very motive forces that propel humanity, and that when these seemingly negative elements impose their influence, our "manners are softened, the human mind becomes more enlightened, and separate nations are brought closer to one another. Finally commercial and political ties unite all parts of the globe."[47]

While selfishness, ambition, and vainglory have not improved our manners nor rendered the human more enlightened and peace-loving at every point in history, those nations that have successfully implemented the central tenets of the Enlightenment – politico-economic liberalism, equality, etc. – show results that often fall in line with this. In a liberal democracy, everyone has rights, and the state is dispositioned to ensure that any one piece of it cannot pull off wrongdoing very easily. Liberal democracies have for this reason managed better than other permutations of statesmanship in the prevention of cases where one individual or group threatens the welfare of others. Economic freedom and schemes for the redistribution of income have enabled selfishness, ambition, and vainglory to ripen the fruit of material progress in such a way that the harvest does not just land in the hands of a few.

The gradual emergence of democracies on the world map starting at the end of the eighteenth century was a moral progression for us. The last half of the twentieth century in particular saw the number of democracies expand as explosively as rabbits do. At the turn of the twentieth century only 20 percent of governments were democratic, while today that share is about half. Democracies for roughly the last one hundred years have also become markedly more inclusive: women and ethnic minority groups are now indisputably equal members of society, and while about two billion people live in authoritarian regimes, almost 80 percent reside together in just one – China.[48]

At any rate, democratization is not the only positive thing to sing about. Harvard psychology professor Steven Pinker asserts that we live in the most peaceful age in all of antiquedom. He argues in *The Better Angels of Our Nature* that instances of war, violence, human cruelty, and bigotry throughout human history have waned.[49]

According to Pinker, one pillar propping up our transition to more agreeable mammals is the birth of the state. Its monopoly on the use of force and its justice structure have reduced violence radically. The so-called humanitarian revolution of the Enlightenment also played a crucial part, spurred on by one critically important innovation: the printing press. People now digested literature, and through the reading of history developed the capacity to identify and empathize with people from other cultures and backgrounds. We merged closer and closer to the notion that every person is worthy, and it is of course harder to discriminate against someone considered to have value than someone not.

Pinker emphasizes how different social circumstances favor the display of different faces of our nature. The achievement of moral progress is thus primarily verifiable in the code with which we are programmed. We have our institutions to thank for chaperoning us into civility, but certain technologies, like the written word, also hacked us a path.

Paying for It Now

It is therefore permissible to conclude that the last three hundred years have indeed produced human progress, but convincing evidence for the contrary has also accumulated along the way.

The last century wasn't an exceptionally bloody period in our history, but it must have at least felt that way to Europeans that experienced its first half. World wars that took mothers from sons and sons from mothers put the idea of a civilized Europe on ice and a big dent in the very suggestion of progress. It's not that the human had become more sinister. It's just that the technologies were now better at obliterating things. The commission of atrocity had never been this easy or efficient.

Technology lengthened the distance between the trigger-puller and the target. People didn't have to be born heartless and immune to savagery in order to do heartless and savage things. Machine guns stitched up the enemy tens or hundreds of meters away without having to dismember the opposing soldier by hand. It was easier to empty the bomb bay onto the head of a pregnant woman from the sky than have to come eye to eye with her and do the deed with a knife. This gradual separation between parties to conflict has in recent years been taken to an extreme. American military men and women have conducted strikes in Afghanistan and Pakistan with the convenience of remote control while tucked comfortably into air-conditioned office quarters on bases stateside.

The immensity of the battlefield bloodbath and the pummeling of cities into rubble with airstrikes in both world wars should already in itself be a crippling blow to progress. But the most reverberating feats of the twentieth century were our forays into nuclear munitions and genocide, both of which are rooted in progress. The atomic bombs employed at Hiroshima and Nagasaki, as well as the hydrogen bomb known to be capable of eradicating nearly all life, symbolized the pinnacle of what science and technology can do at their finest.

Sociologist Zygmunt Bauman has pointed out that The Holocaust was a logical consequence of modernization – the hallmark of the rational planning and effective implementation of a scientifically and technologically advanced society – despite its objectives being irrational and immoral.[50]

The approach of the end of the twentieth century further deepened the doubt about man's ability to "cultivate and care" for the world around him. Material progress enhanced well-being, but in so doing polluted the atmosphere, spoiled the waterways, ravaged the forests, and sent several animal species into extinction. Rachel Carson's *Silent Spring* put momentum into the conservation movement in the United States in the 1960s. It shed light on the harmful effects that pesticides pose to both the environment and humans. Alarm over the use of CFCs (freon) in refrigerators grew in the 1970s and 1980s on account of the rapidity with which they were punching holes in the earth's ozone layer. It was an exceptional case because it was ultimately resolved by political decision. The global agreement signed in Montreal forced states to limit the use of CFCs and eventually replace them almost entirely with other agents.

At the same time, the planet began to take up the task of coming to terms with global warming – or climate change, in the vernacular of today. The resolution of this issue was not going to go as smoothly, owing to the fact that CO_2-emitting fossil fuels cannot be just substituted with something as easily as CFCs. Despite multiple international protocols from Kyoto to Paris calling for caps, the consensus among most experts is that current measures are insufficient to save the world from the melting glaciers, the rise in mean sea level, the drowning metropoles, the droughts, and the famine.[51]

Geologists are preparing to christen a new ecological epoch

with the name Athropocene in honor of the collective stress humanity has imposed upon the biosphere. Humans produce large quantities of "technofossils," an example being the previously mentioned pacemaker of indoor plumbing – the porcelain toilet seat – which is of a construction that promises to last practically forever. Another is plastic trash, in which our oceans are already saturated, motivating one estimate to predict that by 2050 there will be more of it in the sea than fish. So here again the high priests of progress can chalk one up against nature, recording an even more impressive achievement on humanity's CV than the demise of other species about a thousand times faster than their natural extinction rate.[52]

Computers (addressed in the following chapters) are usually immortalized as the messiah of the environment, because they enable telework and prevent the unnecessary waste of paper. But the carbon footprint of just our datacenters is already in the same class as the airline industry, about 2 percent of all carbon dioxide emitted. Every single photograph and Like uploaded to Facebook requires storage space, meaning every photo and Like burns our natural resources. Some analysts predict that the energy consumption of our data centers will triple by the end of the 2020s. The electricity eaten just by IT is said to have already reached roughly a tenth of the entire world's power needs.[53]

Casualties of Our Own Success

Are there limits to social or material progress? Does perpetual advancement make us happier? Can it turn on itself? History provides a long list of examples of communities that fell victim to their own achievements. Canadian writer Ronald Wright refers to this reversal as a *progress trap*. Example: moderate development in hunting tools and techniques during the Stone Age initially led to better nutrition, but more radical advancement caused the exhaustion of local food supplies and starvation.[54]

Events adhering logically to this theme befell humans again after we got into agriculture. When new technology ups the stakes, things can spin into the drink. The Sumerians (~4000-2000 BC), one of the earth's first civilizations, flourished within what is present day Iraq, thanks in part to technological innovation: their revolutionary irrigation network turned the entire region green for hundreds of years.

That same irrigation network, however, was destroying their whole existence. Diverted river water absorbed salts from rock, and after being routed to dry fields, naturally evaporated, depositing its dissolved salts in the soil. Centuries of land salinization eventually made the area untillable. The wasteland sprawling across Southern Iraq today stands as a memorial to the Sumerian irrigation system – a step forward that boomeranged back to bite us in the ass.[55]

Wright applies the concept of the progress trap solely to the environment and ecology, and calls climate change the most meaningful progress trap of our time. The Industrial Revolution improved our lives. Industrialization gave us the stuff we wanted, and modern transport whisked us from place to place. Simultaneously, these good things – industry and transport – (as a matter of general scientific consensus) are destroying the world.

We can of course at this stage sit back and bask in the hindsight of not hunting into extinction the species you need, the imbecile Sumerians for not filtering salt, how we know (or at least think we know) what fossil fuels do to climate, and how prodigiously schooled we are at keeping all bad things at bay.

But how intimately are we familiar with the effects of these newer, more elusive and intricate technologies? Do we really know what we are doing when we develop AI with an intelligence on par with the human? Do we have our hands around every possible consequence of nanotechnology's manipulation of nature at the level of atoms? What about the alteration of human genes? The peak of man's achievements in knowledge and capability is becoming so parabolic it risks tipping into our most nightmarish failure.

It should no longer be a question of *just* the few hundred thousand that might die of malnutrition in the process of some ancient civilization expiring. Many of the most preeminent minds in natural science from Stephen Hawking to Martin Rees have elucidated how one misstep can spell the end for all of humanity.[56] No longer can we chip innocently away at our problems on the playground of trial and error when AI with far more brains than us might annihilate everything (we will revisit this in Chapter 7).

Even if we were more fortunate than ever and new technologies never threatened apocalypse, their intent could still turn on itself eventually. This is fast becoming a reality, as the relentless infestation of digital devices robs our lives of meaning. Endless attempts to build convenience and ease into life do not necessarily enhance well-being.

The idea of the progress trap is not intuitive. We find unobjectionable the thought that good things come from good, and bad from bad. So wasn't scientific and technological advancement supposed to be good? What could possibly be bad about making life easier and spreading amusement? Too much of a good thing can often bring about bad.

Is a Lot Too Much?

The most coveted careers in our society are those in the entertainment world, but the depression suffered by Robin Williams, Robbie Williams, and many other celebrities illustrates how stardom and adoration can breed unhappiness. Those who have achieved and seen everything have nothing to shoot for, and life can easily turn meaningless. In the words of Oscar Wilde: there are only two tragedies in life – one is not getting what one wants, and the other is getting it.

Sofia Coppola's film *Somewhere* throws its focus on the emptiness of those who meet with too much success. The film's protagonist – a famous actor – is drenched in money, women, and status. He is admired, his life is free of inconvenience, but not only is he unable to enjoy his women and sports cars, the sight of live strippers puts him to sleep.

At the level of society, we run about our lives in a similar sphere of stagnant air. Where are the dreams – the big projects whereby we can make our society better and our lives more meaningful? What party can say the political process and material progress can still yield something fundamentally more valuable for us? A hundred years ago, the situation was different; today the forward march of economic-technological advancement threatens to take from us our jobs, our source of income, and our purpose (more on this in chapters 5 and 6).

Happiness research does not corroborate the line about economic-technological advancement making us happier. Economist Richard Easterlin observed in his classic 1970s study that happiness in America had not increased after the Second World War despite the economy formidably growing. But the thing that makes these statistics somewhat ambiguous is the fact that the rich are on average found to be happier than those who are impoverished. Citizens of rich nations are on average slightly happier than residents of poor. This inconsistency is called the Easterlin Paradox.[57]

The main explanation behind the happiness reported among those of great wealth is that they are sustained by an ability to win the war of status. Economic success at the individual level, however, doesn't increase one's sense of well-being indefinitely. Recipients of the Noble Prize in Economics Daniel Kahneman and Angus Deaton have shown that after penetrating into the upper middle class, more money usually does not connect with more happiness.[58]

Being human means being susceptible to the tendency to

compare our lot with those around us. This was being done already during the time of the Roman Empire. The question of how much money is wafting around the economy in such and such an era is secondary to the question of how well off we are *compared to others*. Being aware of this relative metric expands the self-satisfaction of citizens in rich nations as well as individuals receiving stratospheric incomes. The other side of it is less effervescent: while the poor in Europe today enjoy far better material conditions than they did in the 1970s, that fact doesn't make them happier. Their point of comparison is their contemporary – not the poor of the 1970s (on the other hand, many other things besides money, like human relationships, might make them happier).[59]

I am not proposing here that the removal of scarcity and strife from society would not have improved life. Economist and long-time happiness researcher Richard Layard admits that happiness in society increases in parallel with progress, but not indefinitely. As soon as GDP per capita exceeds about 20,000 USD per year, further growth typically fails to add to society's happiness or alleviate its social problems. Rich nations in the West largely crossed this threshold in the 1980s.[60]

A few years ago, a study emerged that found that Finns would prefer to live in the 1980s more than any other period. Pundits lambasted the public for not thinking critically. Show host Tuomas Enbuske tried to remind us how in the 1980s there were no mobile phones or worldwide web. Bingo. Perhaps this is precisely why that period tugs at us with such nostalgia. In the 1980s, significantly more time than today was devoted to being socially interactive.[61]

It is of course somewhat of a subjective assessment to try to draw a line beyond which further material progress is not necessary. Some look for answers in the past. Rousseau's ideal was the mythic and nature-venerating Arcadia, somewhere between his own age (the 1700s) and hunter-gatherers. There you could leave behind "your restless minds, your corrupted hearts, and your frantic desires." Some people might prefer to pause progress during their childhood or youth. Everything that comes later proves more cumbersome to stomach, having grown up within the confines of another reality.[62] As for me, I would have stopped the clock in my childhood – in the 1980s or 1990s. At that time, you could already watch He-Man on TV, but the invention of imaginary friends was still unaffected by the invasion of digital devices.

/ / /

Technological advancement reduces human input and frees us from a life beholden to hardship. There is much to appreciate about this. Aristotle wrote that the arrival of machines (automata) would mean that slaves would no longer be needed.[63] But when taken to its extreme, progress makes us lazy, unimaginative, and incompetent. We forget that we enjoy more thoroughly those things that demand that we exert ourselves. We seek feelings of achievement and reward. The modus operandi of the engineer – IT or not – is instead centered on the presumption that people just want things to be easy and delightful.

Apropos of this still are the remarks of the late environmental activist Pentti Linkola from 3 decades ago:

> All I do is just give the cord a little yank, press the button or turn the lever, and then I just sit, and the banging, rattling, and slamming gadget takes me the length of a canoe, then quits, but it still takes me there, and all I do is sit...The noise, the smoke, the power! Isn't man genius and life grand!!! I don't have to use my arms, my thighs, my back, or any other bit of this primitive jumble of muscle and tendons that nature gave me for no reason![64]

The internal combustion engine made it possible to take the human body out of the equation. Digital devices now open this road for our brains. From here, we take a look at the flip side of IT.

3. Brain Drain

How digital devices destroy thought, dull intellectual culture, and threaten democracy

Where is the Life we have lost in living?
Where is the wisdom we have lost in knowledge?
Where is the knowledge we have lost in information?
TS Elliot (1934)[65]

There's too much sensationalism, misinformation and polarization in the world today.
Mark Zuckerberg on Facebook, 19 Jan 2018, 22:36

I'm a cheap catch. For the last 2 hours I've sat enamored in front of the net, the control center of my orbitofrontal cortex unwilling to fight back as advertised. Self-hate for my wandering focus serenades my utterly futile efforts to get anything accomplished.

And what path did I take to get here? First, I got lost in the froth of Whitney Simmons' fitness YouTube channel, then checked the inboxes of all three of my email accounts, but my hippocampus is no longer responding to inquiries about what followed.

The Siren in my subconscious lied about how allowing myself to trod aimlessly through the internet might actually make me an expert on how being connected inhibits concentration, rewards us for not thinking, and whittles the whole intellectual substance of culture into idiocy (hopefully, this chapter is not evidence of that). But even absent such an excuse, I would have rustled up some other disclaimer – regardless of the fact that I consider myself generally better than most at staying on task and circumventing those on the net trying to suck me into something (like how 90 percent of drivers consider themselves better than average at driving).

If I were to pound this into an old typewriter, I might have more faith in my ability to focus on the real meat of things. But today, with just a few button clicks I have at my fingertips all the world's shopping centers, casinos, exquisite women (clothed and not), immediate access to acquaintances and complete strangers, as well as every movie and song ever thrust upon the world.

The black box that is the internet pulls its fatigued victims in unless actively resisted. Leaving oneself to wander aimlessly about the web is perhaps the digital age's way of getting lost and winding up in a bar in the middle of the workday, but this watering hole exists within computers and weakens the will power of even those who aren't binge drinkers.

Nicholas Carr writes in his Pulitzer prize-nominated *The Shallows* that the internet is predisposed to serve as a place of interruption and interference. Within the web, people are capable of concentrating on one thing for a noticeably shorter period than when reading physical text.[66]

We react to external stimuli as a result of our evolutionary upbringing. As a matter of survival, it has remained important that we take an interest in abnormal noise in our vicinity and minute movements on the horizon. Susceptibility to a shifting focus thus saved our hunter-gatherer forebears from danger. We react with the same level of vigilance to internet notifications that we once paid to attacking lions.[67]

In addition to our evolutionary makeup, the prevailing economic order also vies for our attention. The remuneration model of most internet business is founded not on us being able to concentrate, but rather on the premise that we haphazardly click on everything. In order to more totally rob us of our focus, companies have the world's most renowned psychologists on their rosters, who are plainly aware of what is being siphoned out of our psyche. Former Googler Tristan Harris has stated "there's a thousand people on the other side of the screen whose job is to break down whatever responsibility I can maintain."[68]

Capitalism's battle for the high ground is now fought in the most intimate of settings – the mind. When the former wins, the latter loses, which in the long run is not even sustainable economically – a point we will get into later in the chapter.

An inability to focus is not just my problem. Research shows that the internet environment and smartphones significantly undermine our ability to execute tasks that require highly developed cognition.[69] Digital devices encourage multitasking – jumping from page to page, or the use of multiple apps at once. Multitasking has been linked to wandering and the degeneration of concentration in our primary duties.

Jacking around on your phone while trying to get something done can delay its completion by up to 400 percent.[70] If someone in the same room as you is engaged in multitasking, your own ability to focus is inherently degraded.[71] Negative feelings have been shown to motivate an attention shift elsewhere.[72] We seek a sort of solace from the limitless expanse of the greater digiscape. Boredom becomes even harder to bear today when we are swimming in interesting alternatives.

British research has found that smartphones are checked about once every 4 minutes and 20 seconds – an average of 221 times per day. Try to concentrate on something while doing that. In fact, evidence suggests that smartphones are a disturbance even if not glanced at. A 2015 study found that exposure to just smartphone notifications generated underperformance in intensive tasks, despite test subjects not even reading the alerts. A separate study revealed that just a smartphone (yours or not) sitting on the table lowered scores in tasks that require cognitive performance. It has been thoroughly regurgitated that technology in itself is neither good nor bad, and that its success just depends on how one uses it. But one need not even be using it for its mere presence to pose negative effects (we will touch on the theory of technology's neutrality later in this chapter).[73]

What this means is that we cannot focus when we are with digital devices as well as we do when we are without them. But does the influence of our smartphones persist even if we do not bring them along? Research has yet to definitively establish that digital-device use makes our brains more vulnerable to being interfered with. Verification of this question would require many years of extended observation, something brain health specialist Mona Moisala says will remain hampered by the difficulty to secure funding.[74]

Moisala's dissertation work nevertheless suggests a connection: those who frequently multitask online are less able to concentrate than others. A separate study uncovered how multitaskers have less gray matter in areas of the brain responsible for controlling attention. There is also evidence pointing toward the weakening of both their long- and short-term memories. These studies are not proof that multitasking weakens our ability to concentrate. It could just be that those naturally unequipped with a facility for focus prefer multitasking.[75]

It is clear, though, that the online skirmish for our attention is brutal, and that our interest is more easily corralled by the exciting, the visual, and the urgent – none of which can slide by without in some way influencing how we ingest stimulant. "If we busy ourselves each day with just snippetry – which social media implores we do – then it is reasonable to expect that it is eroding our ability to concentrate for long periods."[76]

The question of whether or not these changes to our brains are permanent is of secondary importance to the fact that we are glued to these machines, and they are not going away anytime soon. Just trying to concentrate in close proximity to them is arduous.

Digital devices poison the joy that comes from self-discipline and delayed gratification, which are indispensable components of social order. In a special experiment at Israel's Ben-Gurion University, smartphones were distributed to people who had never used them. Impulsiveness mushroomed among test subjects after 3 months of use. The mental steadfastness required by delayed gratification lasted far less than it did previously, despite the promise of a larger reward later. Participants instead selected smaller rewards for which they didn't have to put up such a struggle (their ability to process information was also shown to have weakened).[77]

There is a growing body of evidence for the ongoing fight to the death over our attention – one that doesn't emerge in formal research. This tragicomedy is told by the proliferation of apps designed to limit internet use against the will of the person using it. In the Foreword to her novel *NW*, prize-winning British author Zadie Smith thanks Freedom© and SelfControl© for "creating the time" for her to write. Both apps prohibit access to certain webpages, while Freedom© can be set to block access to the net itself. (what more could be inferred from the name?). Another well-known British author, Nick Hornby, has also praised the software for having helped him.[78]

And that's not all. Shopping malls have tried to sell products in which you can jam all your electronic thingamajigs that – try as one might – cannot be opened for an hour after the lid is closed (except with the emergency switch, which renders it thereafter unusable).[79]

Glad Someone Remembers

On top of our concentration, our capacity to just remember things can sublimate away in cyberspace. When Google predicts your search word before you even finish typing it in fully, there is no need to rack your brain anymore about how to spell Friedrich Nietzsche. Why perspire trying to onboard this drivel when you can just look it up online?

An essential piece of learning is the recollection of

information – the transfer of it from long-term memory to short. It has been observed that students remember less of a subject when they pass their study time reading than those who try to recall as much as possible of what they read without the book. If we always seek the safe haven of Google instead of attempting to remember the issue in question, both our memory and our ability to learn can be compromised.[80]

Columbia University psychologist Betsy Sparrow has put forth the notion of a so-called Google effect. Participants in one of Sparrow's widely cited studies were asked to type various tidbits of trivia they had been taught into a computer. Some were told that their inputs would be erased; the rest were told that they would have access to them later. After being asked to memorize what they were inputting, those informed that their data would be saved displayed that they remembered it worse.[81]

Fairfield University professor Linda Henkel discovered that people recall art exhibits from memory more poorly after snapping photographs of them than if they had foregone the cinematography and just examined them in detail (on the other hand, those who zeroed in on some particular aspect of the exhibit while taking a photo remembered the work better than those who took no pictures).[82]

When the information we seek can be found with just the flick of a screen, we are absolved from having to go to any length to remember it. At the same time, more and more of our long-term memory gets devoted to chores it wasn't meant to be tasked with. Sparrow says we spend more time today remembering where to find things instead of remembering them.

It is worth noting that our long-term memory is not just a warehouse to shove our stuff when not using it. From time to time we hear it said that because in the age of the smartphone we do not have to remember anything, our brains now have more space to be creative. But memory is a central member of both thinking and intelligence, and cannot be outsourced without confronting repercussions. Memory plays a leading role in the construction of schemas that support analysis and the comprehension of complicated things.[83] Google can of course come in quite handy in digging up blips of information, but if we divest ourselves of the act of remembering, no amount of Googling will ever teach us anything. If like me you cannot recall one iota of Margaret Mahler's theory of childhood development, it would be difficult to piece together something fresh from it into an opinion. First and foremost, we remember – in order to understand and then create anew. Subcontracting the act of recall to Google like this detracts from autonomous and inventive thinking.

Memory is an indivisible fiber in the fabric of our identity. What would we become if a growing portion of our individual history – the building blocks of us – were not present in any corner of our consciousness? Well, thankfully we have those thousands of pics and vids wafting about in hard drives and the cloud, so we can always check to see what went down and who did it.

People Err, Machines Only Get Better

Blindly operating at the whim of machines corrupts our knowledge and our capabilities; we become less relevant when our alacrity for being critical is left to wilt. At issue is not just us amateurs, but rather professionals in positions of responsibility that put far too much faith in their computers compared to their own aptitude.

The realm of the modern aircraft cockpit is one example where this has become prevalent. The US Federal Aviation Administration (FAA) issued a Safety Alert in 2013 encouraging operators to promote manual flying operations when appropriate, because trusting the autoflight systems too much can lead to a "degradation of the pilot's ability to quickly recover the aircraft from an undesired state."[84]

When an Air France flight to Paris from Rio de Janeiro

crashed in the Atlantic in May of 2009, the lives of all 228 people aboard went with it. Investigators characterized the incident as stemming from a gross operational error on the part of the flight crew following the disengagement of the autopilot after losing speed data from the aircraft's frozen pitot tubes. The investigation concluded that the pilots displayed a "total loss of cognitive control." An analysis offering essentially the same conclusion had come just a few short months prior when a Continental Connection flight aerodynamically stalled over residential Buffalo and ended the lives of 50 people.[85]

Reliance on GPS is not news. A charter bus nearly 12 feet high rammed into a 9-foot-high concrete tunnel overpass in Seattle in 2008 due to the driver stubbornly refusing to prioritize his brain over digital route guidance. He said he "didn't see" the flashing lights and signs warning him about the low height of the tunnel. Twenty-one children were injured. Americans have driven straight into homes, and Australians right into the sea – all because GPS commanded that it be done. There are several examples of cruise ships that have gone adrift or run aground because their deck officers couldn't believe that their integrated bridge equipment could be in error.[86]

The quiet crumble of our critical faculties is also emanating from within the media. Few and far between today are hosts with a grip on independently aware journalism. The yard stick of a good story now is more a matter of how many shares it gets on Facebook. Criticism is no longer in their job descriptions, having been jettisoned for the data stream and whatever the masses think. Research indicates that the decay IT brings to our critical competencies is conspicuous in many other educated professions, from specialized medicine to architecture.[87]

This abysmal state of affairs is also reflected in numerous apps. Should you carelessly leave your trousers awry, Noti-Fly will promptly direct your attention to it. Smart Tampon will send your phone a chime when it has decided that it must be replaced. Lucy Kellaway has aptly disparaged these valueless apps, saying:

> the fact that people are so willing to pay over the odds for non-solutions to non-problems is the best evidence of the irrationality of the consumer the market has yet provided. If we want smart gadgets, we must be dumb...If we no longer have to remember to do up our flies, or look at the sky before putting the washing out, and if our favorite conversation is who walked or brushed their teeth for longest, it's our brains that I fear for.[88]

For Want of Chalkboards

Can the tyranny of digitalization eat away at our performance in school, as well as our intelligence? In my home country of Finland – where the newest technology is always sought after – high school students recorded the highest scores on the international PISA test in 2006, but beginning in the next round and extending through 2018, the figures have decreased.[89]

Apple's iPhone came to market in 2007. Countless studies have subsequently shown that significant social media and smartphone use is connected to lower test scores. [90] Measures limiting student mobile phone use have also been correlated with improved performance in the classroom – most notably among those who aren't normally over-achievers. Researchers from the London School of Economics write that "banning mobile phones could be a low-cost way for schools to reduce educational inequality." France outlawed classroom smartphone use in autumn 2018.[91]

Test scores in many other technologically developed nations besides Finland have taken a dive over the preceding decade. SAT assessments of critical reading, writing, and math skills in the US have gone down since 2006.[92]

Reading comprehension is not a talent that spontaneously

wills itself into existence, but rather something that must be taught and fostered in the right environment. If reading is more and more the act of superficially skimming the surface of the internet, the risk is that less and less of us learn to grasp what we read and to analyze it critically.[93]

IQ tests paint a similarly vivid picture of our illiteracy. The average IQ scores throughout the entire twentieth century rose across the globe by approximately 3 points every 10 years. This phenomenon is referred to as the Flynn effect. In many countries, however, growth has stopped, with IQ figures in Finland, Norway, and Denmark starting to decrease halfway through the 1990s (as test scores in the US have mainly risen). James Flynn and Michael Shayer compiled intelligence research from various nations and observed that it is precisely in Scandinavia where the IQ numbers are dropping most rapidly. In Finland, they decrease by about 0.25 points per year, or 7.49 points in a generation (30 years). Shayer attributes this shift to what he calls a "large social force," an allusion to the bloated role new technology plays in the life of a child.[94]

At the same time, there ought to be less impasse now for success in school and rising intelligence. For example, leaded fuel use brought about developmental disorders for decades and weakened school performance – but these were banned in the Western world by the mid-1990s. Young people run to alcohol now for intoxication less than previous generations, and thanks to the internet, information ought to be readily available to everyone.

But of course we shouldn't read too far into the downward slope of test scores, as there are a slew of factors other than digital devices at play. One theory posits that our descending intelligence level is due to greater numbers of intelligent women just deciding to not have kids.[95]

Student interest in the natural sciences has indeed dipped in some countries, and this may in part explain the drop in PISA results.[96]

Either way, the sparkling starlight of digitalization has been horribly overstated. China's Shanghai and Hong Kong stand on the top of the PISA score pyramid, but they sit in the bilge when it comes to IT use in the classroom. The economic forum for rich nations, the OECD, stated in 2015 that despite sizable investments in information technology, test scores at schools have not improved.[97]

Students making moderate use of IT at school get slightly better grades on average than those who make rare use of it, but students who heavily use IT in school record markedly below-average scores on almost everything. One interpretation for this presented by OECD researchers is that "building deep, conceptual understanding and higher-order thinking requires intensive teacher-student interactions, and technology sometimes distracts from this valuable human engagement."[98]

It is noted in this exhaustive study's Foreword that:

> perhaps the most disappointing finding of the report is that technology is of little help in bridging the skills divide between advantaged and disadvantaged students. Put simply, ensuring that every child attains a baseline level of proficiency in reading and mathematics seems to do more to create equal opportunities in a digital world than can be achieved by expanding or subsidizing access to high-tech devices and services.[99]

It is worth mentioning that the innovators of our newest technologies are often wary of exposing their progeny to these inventions. The late Steve Jobs of Apple, as well as Bill Gates, the founder of Microsoft, imposed strict limits on their children's use of technology. Instead of fiddling with phones at the dinner table, the Jobs family conversed about books and history. The offspring of Silicon Valley's heavy hitters regularly attend schools where new tech is nowhere to be seen. At the Waldorf School near Google's headquarters, for instance, students continue to use chalkboards and pencils.[100]

Chris Anderson was editor-in-chief of emerging-tech magazine *WIRED*, and today serves as the CEO of drone manufacturer 3D Robotics. He told *The New York Times* point blank that he forbids his children to use smartphones.

"My kids accuse me and my wife of being fascists and overly concerned about tech, and they say that none of their friends have the same rules. That's because we have seen the dangers of technology firsthand. I've seen it in myself, I don't want to see that happen to my kids." By dangers, Anderson is above all referring to device-addiction and internet filth.[101]

We Are Playdough

Whenever a discussion of technology's dark side arises, the digienthusiast pipes up about how devices themselves are not bad – and that the question is rather one of method of use. This tune is strum with egregious frequency, but incessantly repeating it does not make it any truer.

We are more beholden to circumstance than we care to admit. Social philosophers beginning from Hegel and Marx have illuminated how our awareness is a product of our social reality. Numerous social psychology studies and related findings show that our environment shapes our nature. Stanley Milgram's early 1960s experiments on obedience revealed that under authoritarian pressure, good and law-abiding people were eager to administer deadly doses of electricity to actors screaming in agony in adjacent rooms. Analysis of airline hijacking incidents made it clear how otherwise altruistic, married, and family men in the middle of a crisis can do whatever it takes to save only their own skin. Loving mothers interned in concentration camps have discarded children in order to ensure their own survival.[102]

Although they are on the extreme end of the spectrum,

these examples show how preposterous it would be to insist that people always conduct themselves heroically in the face of imperatives. It would be absurd for us to assume that the focus-destroying effects of digital devices tell a tale of individual inadequacy and weakness – and not one of conditions they create. It would be hairbrained of us to think that the inane solution to the whole matter is for us to just purge ourselves of our impurities.

In the 1960s, media theorist Marshall McLuhan referred to the traditional view that it is how media and technology are used that counts as the "numb stance of the technological idiot." He drove the point home with a rag about how one could just as well say that the smallpox virus is neither good nor bad – it is the way it is used that determines its value. To McLuhan, our various technologies represent "extensions" of our bodies. They control the course of our interactions and unobtrusively reformat us into new beliefs. "The effects of technology do not occur at the level of opinion or concepts, but alter sense ratios or patterns of perception steadily and without any resistance."[103]

McLuhan died in 1980 without ever seeing the internet in its modern form, or the power with which it plies us into shape. It makes waves in the way we read, think, and communicate, and it dunks our whole culture in change.

Oxford neuroscientist Susan Greenfield warns that the detrimental hold technology has on our brains is a tectonic shift no less disturbing than climate change. She loathes the use of lingo in which technology's adverse effects can be prevented by simply using them in the "right" and "rational" way. "When has common sense ever automatically prevailed over easy, profitable, or enjoyable possibilities?" Greenfield asks, while reminding us of how the demise of 100 million people from smoking over the last century does not exactly underscore man's ration.[104]

It should be noted, however, that the written word was

eyed with foreboding suspicion even by the ancients. Socrates maintained that writing will "create forgetfulness," because those who practice it "will not use their memories." He said literary culture will offer the potential for "having the show of wisdom without the reality."[105]

It is highly probable that prior to the birth of written literature our ability to remember things was much better, but the fact remains that its emergence generated volumes of added value. It spawned an expansion of perspective from one's own village and immediate community into the extreme corners of the earth. This is what made science and knowledge available to more people, and enabled us to develop a (relatively) objective sense of our world.

Internet culture promotes the spread of information at a speed unattainable by books, but these two cultures – literary and digital – have one fundamental difference. Literary culture is an intrinsically silent and introspectively oriented act that cultivates reflective, independent, and uninterrupted thought. It fuels the development of imagination, the creation of vibrant inner worlds, and discourse based on painstakingly laid foundations. The ethos of the digital age is instead a social spiral of reaction and reward, the instant fulfillment of pleasure, and the repackaging of everyone's emptiness as sincerity.

More succinctly generalized, literary culture produces quiet, considerate, precise, and methodical human beings, while the digisphere turns us into over-social loudmouthed wankers lapping at screens.

In this new environment, instead of our own volition, we operate more on the terms of our contraptions. Henry David Thoreau wrote in the 1800s about how man has a tendency to become the tool of his tools.[106] Few of us want to drain away our days reacting to smartphone notifications from Facebook, Twitter, Instagram, and others. But still, that is where most of them go. We are the bullseye of our own bombardment, and it is to this well-aimed digital artillery barrage that we have surrendered.

I do not mean to impart the conception that being reflective and circumspect with respect to the web is impossible – I am just saying that the internet world pulls us in the complete opposite direction. Anyone in a position of undisputed power can still choose to rule with some semblance of ethics, and in a concentration camp one can still resist the urge to just save oneself. But in all of it, the environment is yanking us the other way.

I am neither saying that the focus-loving culture this chapter beautifies is some pristine piece of nature that the internet is polluting. The age of literary culture is but a thin sliver in our history. Evolution has made us into beings for whom distraction will remain a contagion. It is quite a separate question then – do we want to construct our culture on the basis of us staying barbarians?

The Age of the Less Than 17,079-Page Novel

The reading of print media has begun to get buried as the art of a bygone era. The time we devote to reading has dwindled in recent decades, even when the hours spent reading on the net are included. This decrease is particularly visible among people aged 15-44.[107]

Websites are digested differently than material on physical paper. Jakob Nielson's investigation into reading habits asked how users read on the web, and it answered by concluding "they don't." The interval we spend loitering on one webpage is typically about 15 seconds. Bounding around from page to page and the implosion of the period in which we read are also themes common to the perusal of the published works of academia. Many researchers find the prospect of reading from the web unappealing.[108]

Nicholas Carr points out that skimming is becoming the reigning way that we read. "Once a means to an end, a way to identify information for deeper study, scanning is becoming an end in itself – our preferred way of gathering and making sense of information of all sorts."[109]

The shrunken role reading plays in our culture does not just alarm the anachronism that is the Luddite. Google's long-time chairman of the board, Eric Schmidt, is similarly perturbed: "I worry that the level of interrupt, the sort of overwhelming rapidity of information...is in fact affecting deeper thinking. I still believe that sitting down and reading a book is the best way to really learn something. And I worry that we're losing that."[110]

Classics that run on without end such as Kalle Päätalo's *Iijoki* series (a literary fossil in Finland that contains 17,079 pages in total) were treasured during the golden age of literature in part because the culture that accompanied them was not a firehose of mishmash. Today we can rightly ask if good art is exactly that which doesn't survive the ongoing fight over our attention – due to simply being too boring.

In an age of overflowing data, there is cause for us to chomp down on the positive points of a world of minimum stimulus: it forges a more resilient intellect and amplifies the colors of our internal pallet.

Susan Greenfield has stated that "understimulation" – as a counterweight to the overstimulation of the smartphone – is often helpful, particularly in the development of children. "A tree in the park didn't ask you to climb it, a drawing pad didn't ask you to draw on it," she writes. "Sometimes, even quite regularly, you might get bored. But it was that very state of understimulation that impelled you to draw a picture, make up a game, or go outside to play...[then] *you* were the driver and you would be in control of your own inner world, your own private reality."[111]

Metamorphosis of the Megaphone

Internet culture encourages us to express ourselves with photos, videos, and whatever entertains. When the war for attention is never-ending, those who speak to our inner animal win. Kim Kardashian's finely formed fantail and forecastle are enough to keep 68 million Twitter and over 195 million Instagram users in tow. The more than 200 million people following Dwayne "The Rock" Johnson on Instagram care less for what he writes and says than the rippling symmetry of his pecs.

Social media is adept at swindling us out of the time and data that internet businesses seek because it can tap into the aquifers of egocentrism and our need for recognition. The track about which we orbit our own barycenter has been refined by new Facebook algorithms introduced in early 2018 that ensure our Friends' content – pics and status updates – show up at the head of the page. It is of course more pleasing to wake up to a pictorial litany of others' daily drudgery than the misery with which Myanmar's Rohingya minority might be grappling.

The crucial question, though, is what will the movement of the nucleus of value to communication by photo, unofficiated baboonery, and the constant revolving around ourselves do to the culture of discourse and the political process. Our problem, as illustrated by Aldus Huxley in *Brave New World*, is that few of us even give a damn. If people are no longer capable of casting an enlightened and critically informed vote, the substance of democracy becomes at best a quivering question mark. More than 200 years ago, Thomas Paine wrote that "when men yield up the privilege of thinking, the last shadow of liberty quits the horizon."[112]

In a way, we are steaming headlong into the fog of a liberalist paradox. Can freedom of choice, when taken to its extreme, kill freedom? Everyone in a liberal society has the right to be superficial and ingest whatever media they please. But when the potluck becomes just chocolate and cheeseballs, democracy is in cardiac arrest. An old adage says that we the people will get precisely the leader we deserve. Americans got Donald Trump, which isn't necessarily the bottom of the barrel, because with the net empowering our selfish and shallow sides, we likely haven't seen anything yet.

Technology has run the discourse into the rough before. Political discussion prior to the advent of television favored reasoned and grounded (or to us, boring) argument. For example, during the famous debates between Abraham Lincoln and Stephen Douglas in 1858, the first candidate afforded the floor was given an entire hour for opening remarks. His opponent countered for an hour and a half, after which the other was given a half hour to reply. It wasn't a belligerent barking match with the boots of both stuck so far down each other's throats neither could cough up what they wanted to say. Thousands listened to the debates on location with a calm, concentrated, and eager quiet.

In today's media environment, Lincoln – widely believed to be America's most magnificent president – wouldn't have much chance at winning the presidential election. He was an articulate and deliberate speaker, and unfit with the speed needed to successfully one-up every talking head. He spoke in a high register, and he wasn't very handsome (height notwithstanding), but in his time looks did not matter. The majority of those in attendance could barely even see, let alone analyze, the fashion of the distant figures on stage.[113]Beginning in the 1960s, televised debates significantly changed the way in which politicians were presented. It became more essential than ever to express oneself quickly and flippantly – and, of course, to look good.

Today the giving and taking of the floor for the politician often occurs virtually on social media, where more meaningful than the profundity or plausibility of an argument is the dispatch of short, pilfered, and belittling quips.

Technical meddling within the system helps land punches when spoken words alone no longer work. Twitter visibility can be made more formidable with various autonomous and semiautonomous bots that spoof its algorithms. This trickery is more or less what determines which tweets show up in front of everyone first, because Twitter's algorithms identify those tweets that are "most popular" and prioritize them in everyone's queue.[114]

The tray of most popular tweets is therefore rarely anything but hype – the programmed output of an unfeeling machine, and far more than the Russians poking around in Western politics. Those same lines of code are welcomed just as warmly by the world's democracies because they magnify the mouthpiece of both the right and the left. Info war expert Molly McKew says of the Twitter conversation that, "[H]uman minds are on a battlefield between warring AIs."[115]

The political shoutingscape hasn't just become shallower – it is now a debacle waged in the absence of both umpires and rules. In the age of the social media uproar, the politician is susceptible to pressure to jump into the general fray. Few dare to wade into the data stream and take a stand on behalf of something different.

I Don't Wanna Speak with Them

In the mid-1990s, MIT tech pioneer Nicholas Negroponte predicted that soon everyone would be devouring news tailored to their personal preferences. He believed that it would no longer be necessary in the future to stick with newspaper and television, and that our "daily me" would be hemmed together from multiple sources that tickle our interest.[116]

Wait no more. Facebook, Twitter, and various news apps have sealed us in our own private cocoons. We can choose to listen to, watch, and read only those opinions that entice us (if such things entice), but as we know, cocoons ensure we only have half the facts – and that they are distorted.

Legal scholar and former advisor to the Obama administration Cass Sunstein warns in his book *#republic* that info cocoons are a danger to the whole political order. Sunstein emphasizes that freedom of expression is not defined just by the absence of censorship, but that for freedom of speech and democracy to function, two other requirements must be met. First, people must be exposed to content (and people) that they didn't select ahead of time. John Stuart Mill wrote in the mid-1800s that "it's hardly possible to overstate the value...of placing human beings in contact with other persons dissimilar to themselves, and with modes of thought and action unlike those with which they are familiar." What we need are chance encounters. We should be astounded by each other, and perhaps even annoyed. Such a process is uncomfortable, but it can change the way we see things – and bring us closer to the truth.[117]

The second requirement is that citizens have shared experiences, not just those that are shared with other people in the cocoon, but experiences that are shared nationally and even internationally. Without these, it is difficult to both comprehend and address social issues, i.e., engage in the political process.[118]

It is certainly possible to cross paths with differing opinions on social media, but in many cases, they are rabidly extreme. Facebook operates on protocol that runs the most controversial topics through the top of the ticker, because those are the ones that best whip the people into a fit. Likes and Dislikes. Shares because you agree, Shares because you hate. Following the Facebook newsreel can wrap an air of fanaticism around those who just think differently, and this can drive opposing tribes into the trenches – especially when instead of trying to convey perceived discrepancies in another's argument, we focus solely on trying to badmouth everyone on their side. Those who forgo social media for the nightly news or the perusal of the morning paper see the world in a somewhat more harmonious way.

During the 3 years that I hosted *Perjantai* (a Finnish prime-time talk show), it was reaffirmed for me almost every week how twisted our conversational culture has become. Today many come off nearly aghast at the fact that people with different opinions can be gathered in the same studio. Disagreement used to be prerequisite for a good conversation. Now all parties are so polarized none can tolerate a discussion of the actual issue, and just raise a fuss about having to share the same air with the enemy.

Any group amassed under another banner can easily be seen as evil and uncaringly ostracized. Respected academic argument can even be labeled racist, socialist, or otherwise extremist – before any effort is expended to determine what is being said and what is not. As Jesse Singal wrote in *The New York Times*, "It's getting harder and harder to talk about anything controversial online without every single utterance of an opinion immediately being caricatured by opportunistic outrage-mongers." Singal says because of this, everyone "rushes for the safety of their ideological battlements, where they can safely scream out their righteousness in unison."[119]

As the facts about anything are unique to our cocoon, opposing sides are often blamed for disseminating fake news, despite those doing the accusing not standing on any sturdier ground than the accused. Rhetoric in the age of the internet includes the brandishing of participants as trolls for no other reason than that their opinion sits at the other end of the spectrum (I'm not denying that fake news and trolls exist – just pointing out that trading ugly names makes reasonable conversation difficult). In years past, people made a place for disagreement, because we were more often engaged with people who represented something different. Now many of us just switch off and flee for our comfort zones.

The inter-party quibble that is the US political system often comes up with some funny numbers. Americans were polled in 2010 about how unsavory it might seem if one of their children were to wed someone from the opposing party. A total of 49 percent of Republicans and 33 percent of Democrats referred to this as unfortunate, but today those figures are almost certainly higher. Many US employers say they shy away from hiring people of the "wrong" party.[120]

Harvard political science professors Steven Levitsky and Daniel Ziblatt highlight how important it is in a functioning democracy for each camp to adhere to some unwritten democratic norms. These include "the understanding that competing parties accept one another as legitimate rivals," and not enemies. In their book *How Democracies Die* they write that "if one thing is clear from studying breakdowns throughout history, it's that extreme polarization can kill democracies."[121]

Tough talk about democracy's demise might smack of exaggeration, because that is generally held to be the day that tanks are rolling through our cities. But governments in Hungary, Poland, and the United States have thumbed their noses at democratic institutions and declared war on those with opposing opinions. As Levitsky and Ziblatt remind us, the dead democracies littering the ocean floor of history usually sank under the weight of what trickled in slowly without anyone noticing.

For Sale: Traditional Media
Wanted: Fake News and Hate Mongering

Techies love to sling spit at those who long for the newspaper age and broadcast television. The world has moved on, they say, and traditional media houses are history.

Old media, though, contains immeasurable worth, a point obviated in any examination of what gets left in the wake of digital media. As traditional news outlets become the stuff of yesteryear, we look forward to superficiality and polarization. Our future will flower into a smorgasbord of fake news, hate mongering, and centralized data, and all of this for just leaving our smelly relics behind!

There is hardly an entity that has gotten closer to a true

monopoly on information than Google (although many organized religions at one time enjoyed positions that compared to, if not surpassed it). Research has confirmed that the initial results from something punched into the search bar have a tremendous effect on the formulation of opinions. We trust that what comes in at the top of our queries is unquestionably truer than what languishes lower, not to mention that which does not even make it onto the first page. According to the work of Robert Epstein, initial search results are afforded such trustworthiness that their manipulation could change our very voting behavior. About a third of those who live in developed nations use Facebook regularly as their source of news[122] (more on these larger of the gatekeepers, Google and Facebook, in the next chapter).

The uploading of academia onto the internet can bring about a coagulation of scientific knowledge. Research conducted by James A. Evans at The University of Chicago has shed light on how the digitalization of academic publications is causing today's scientific articles to cite a progressively slimmer spoonful of literature. Web searches funnel all researchers to the same crop of sources. Despite the entirety of the world's accumulated knowledge being more easily accessible than ever, rather than distinct ideas shining, our tunneled perspectives could just grow narrower.[123]

No other question of internet culture has been discussed more fervently in previous years than fake news, but the endless recitation of baseless claims should not come as a surprise. Newspaper, radio, and television outlets employ people that are (mostly) dedicated professionals, and decisions about what to publish adhere to editorial discretion and ethics. On social media, any numb nuts can post whatever they please.

This fact is taken as troubling also by those at the top of the tech elite. Former Facebook VP Chamath Palihapitiya has stated that overdosing on the quick dopamine fix of social media is a global pandemic that is "destroying how society works; no civil discourse, no cooperation, misinformation, mistruth."[124]

During the 2016 US presidential election, many shared "news" reports about how the Pope endorsed Donald Trump for president. Others were treated to stories that were also blatantly misshapen, such as the one about how Hillary Clinton runs a pedophilia ring out of its base in a DC pizzeria. It became clear that teens in the small town of Veles in tiny North Macedonia were cooking up many of these stories, and that in so doing were banking some serious cash from advertisements.[125]

Russian disinformation was apparently at work not only in the 2016 US presidential elections, but also in other Western elections, aiming to at least divide the electorate. It's not just about Russia – Trump's campaign wielded the same weapon in its own targeted advertisements.[126]

Fake news in other nations has also led to the instigation of violence. The preferred news source in India is Whatsapp, over which unsubstantiated chatter of child abductions was broadcast repeatedly. At least 30 people were lynched over the course of the rumor mill that resulted.[127]

If Whatsapp is infallible in India, then in Myanmar Facebook is basically another way to say internet. Facebook has been of fine service to Burmese monk Ashin Wirathu and army general Min Aung Hlaing in seeding hate, and of course demonizing Rohingya Muslims with fabrications. Zaw Htay, the spokesman for de facto leader Aung San Suu Kyi, has used Facebook and Twitter to berate the Rohingyas for supposedly burning their own homes. Social media has thus been an incentive and an accessory to cover-up for ethnic cleansing.

The UN itself has extended its criticisms to Facebook for spreading hate, prompting the company to delete several accounts from its platform. Buddhists in Sri Lanka have chosen Facebook as the place to spin fictions about Muslims, leading to cases of people being beaten and murdered. The violent outbreaks in Kenya following its 2017 presidential election were in part propelled by glaringly fake news.[128]

The speed at which disinformation infects people on social media can turn a bad situation worse, but concocted news stories are in no way anything new. The politics game has always involved the defamation of one's opponent and inflammatory statements of untruth. We ought not to forget that disinformation from one side is often met by retaliative disinformation from the other. Russia has mass produced fake news for decades (and the US has meddled in other nations' elections on countless occasions well before the web existed).[129]

As it stands, there are still no strongly definitive data explaining the role that fake news plays in the manufacture of opinion. Three reputable political scientists concluded from data about personal internet use that one in four Americans came across at least one piece of fake news leading up to the 2016 presidential election. Only 1 percent of what Hillary Clinton's supporters read before the election contained fake news, while the number for those behind Donald Trump was 6 percent. The research found that fake news is read most reverently by those that are most stalwart about who they want to vote for.[130]

Oxford researchers, however, discovered that in the battleground state of Michigan, Twitter was more inundated with fake news than real. It is not known how many eventually read or were affected by "news" invented by activists and delivered by bots. A study conducted by The Ohio State University says the influence of fake news in America is nothing less than national. It found that in most cases those who voted for Obama in 2012 and believed at least one fake news piece about Clinton didn't vote for Clinton in 2016, but that nearly every Obama backer that managed to avoid fake news in 2016 voted for Clinton. Using these results, *The Washington Post* calculated that fake news lowered Hillary's total votes by more than 2 percent – in other words, decisively.[131]

Escaping the scourge of fake news cannot be accomplished without introducing new issues. Facebook, Twitter, and Google (owner of Youtube) already enjoy unchallenged preeminence over what shows up in front of us. Today these techno-leviathans are free to decide what does and does not constitute fake (this emerged, one must observe, more out of public pressure than any sort of initiative on the part of the firms). The situation is made extravagantly more problematic by the fact that the quality and quantity of Facebook's, Twitter's, and Google's workforces are wildly insufficient to pull off what they are charged to do. As a result, some of the fake news that was deleted – sites accused of containing support for terrorism, hate speech, or pornography – has been extracted and trashed under corporate hocus pocus without oversight.[132]

/ / /

The felling of the tree of traditional media invites a cacophony of fallacious information to feed on the remains of politics and the republic. Governments should therefore take greater care in taxation and regulation to note traditional media's irreplaceable social value.

Fake news, though, is just one strain in the whole genome of change brought on by digitalization that is shaking the foundations of our intellectual and political culture. As we spend more and more time sloshing around the web, we are reshaping ourselves in its image. The search for shallow and short-lived reward becomes a mind-numbing habit. Meanwhile, focus, prolonged and spontaneous thought, as well as delayed gratification all get steamrolled into oblivion. When constructive intellectual culture stalls, the democratic political order we hold as self-evident is in danger.

A democracy weathers the arrival of fake news from beyond

its borders better than the collapse of its critical resources. Together, they can create one hellishly perfect storm.

4. Every Home's Double Agent

How the harvesting (and exploitation) of big data destroys privacy, due process, and freedom of choice

[T]hey could plug in your wire whenever they wanted to. You had to live – did live, from habit that became instinct – in the assumption that every sound you made was overheard, and, except in darkness, every movement scrutinized.
George Orwell (1949)[133]

You don't realize it, but you are being programmed.
Former Facebook Executive Chamath Palihapitiya (13 Nov 2017)[134]

Great care is usually taken in the selection of those in whom we confide; the most sensitive stuff in us is wide open to the ones we know and trust. Our closest friends thus carry around an enormous capability: they can use what they know against us, because they see where we are weak. It is far harder to be bluffed and manipulated by a complete stranger.

In recent years we have grown so close to our digital devices that we dedicate more time to them than friends. We reveal to machines mountainous amounts of our fears and insecurities, urges and perversions, ailments and affinities, and travels and habits. In this mass of data lies a clear picture of who we are – and the reason we can be easily played with.

Knowing what a person searches for on Google provides a reliable picture of their motivations. Data scientist Seth Stephens-Davidowitz calls Google a truth serum in his book *Everybody Lies*. We feed lines to our spouses, friends, and bosses, but we tell Google our most intimate secrets. Billions of people conduct Google searches, and Google takes good notes. The repository of big data from our search terms builds a realistic and uncensored snapshot of both us and our nature.

A few things can be concluded about humans from their searches: men are often concerned about the size of their penis, women worry about the smell of their vagina, parents fret more over an obese daughter than a son, and people often think as bigots, although being someone of lower economic standing does not appear to mean more visits to racist websites. Stephens-Davidowitz highlights how Google searches are often expressions of feelings rather than attempts to get info ("I hate my boss," "I'm drunk," "My dad hit me").[135]

The world's largest porn site – Pornhub – also trawls along with its net deep in users' search terms. This treasure trove of information explains more about human sexuality than any survey. Pornhub queries reveal, for example, that Sigmund Freud appears to have been right about one thing: fantasizing about the mother figure is commonplace. Some of the most popular porn searches by men contain mother-son iterations, while women are slightly less tempted to seek content starring father-daughter. It might be surprising to learn that the more tender of the genders searches for domination and rape scenes noticeably more than anyone. About a quarter of heterosexual searches by women target females undergoing pain and subjugation. Five percent of what women search for is explicitly rape. Pornhub's data also indicates significant geographical difference in taste. Indian men, for instance, fantasize more than most about being in diapers and drinking breastmilk.[136]

This indiscriminate uptake of data would not be possible without every home's double agent – computers, tablets, and smartphones – and the fact that they follow our every motion. Meaning any hacker that knows what they are doing can have our dirtiest secrets without much of a hassle. The extramarital affairs site Ashley Madison was hacked in 2015 and its user info leaked in an act of public humiliation. Porn sites are also known for selling their search data to third parties (which such a major player as Pornhub would hardly do), and it is always safe to assume that the intelligence community enjoys the same access that site administrators do. America's electronic intelligence arm, the NSA, has dipped into the porn usage of Islamist leaders in an effort to dig up dirt with which to discredit them.[137]

Many might interject here with a statement that there is nothing to worry about as long as you're not engaged in delinquency – no provocative searches on Google and no incriminating posts on social media. There's nothing personal at all about Liking Harley-Davidson on Facebook, right?

Hello nonconformist Kitty

Cambridge University researchers wanted to find out how much of a person can be inferred from just their Facebook Likes. The group, led by psychologist Michal Kosinski, compared available data about a person's Likes with their personality surveys and IQ tests. Findings were published in 2013 after examining over 58,000 Facebook user volunteers in America.[138]

It became clear that with only their Likes, a user's preferred political party can be predicted with 85 percent certainty, their sexual affiliation with 88 percent, and their race with 95 percent. Gay black men who are Democrats Like distinctly different things than straight white Republicans. Relatively accurate assessments of a subject's smoking, drinking, and drug habits were also made solely on the basis of Likes.

Researchers were additionally able to surmise the participants' "big five" personality traits: openness, conscientiousness, extroversion, neuroticism (emotional balance), and agreeableness. It was discovered, for example, that the extent to which an individual is open to new things and experiences is quantifiable from Likes with approximately the same precision as a personality test. Openness, for example, in the American political landscape is clearly correlated with voting Democrat.

Back to Harleys. Researchers took note of how Liking the following Facebook pages is linked to lower IQs: Harley-Davidson, I love being a mom, Sephora, and Lady A. Pages associated with high IQs included the politico-satire *The Colbert Report* and *Science* magazine.

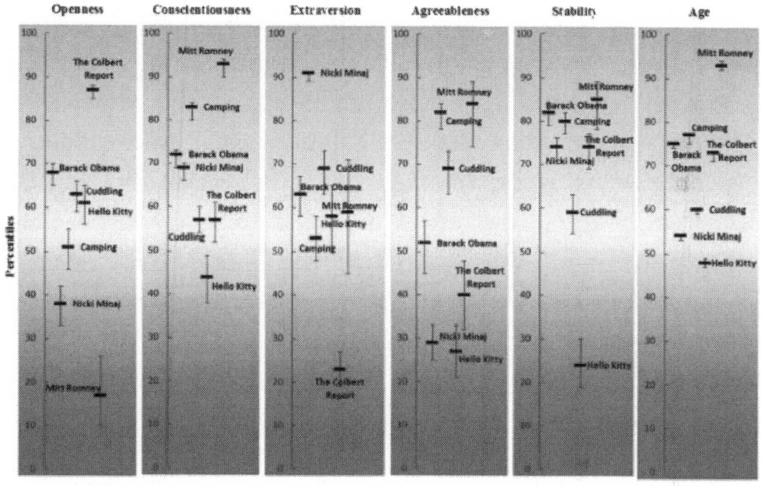

The diagram shows how personality traits and Facebook Likes are associated. From left, Openness, Conscientiousness, Extraversion, Agreeableness, Stability, and Age. For example, those Liking Barack Obama on Facebook are usually extroverted, conscientious, and open to new things. They also exhibit average agreeableness and a high level of emotional stability. Source: Kosinski et al. 2013.

Likes for *Desperate Housewives*, the musical *Wicked*, or Britney Spears point to a user that is gay. What kind of human might harbor a hankering for, say, Hello Kitty? On average, this person is highly neurotic, nonconformist, and open to new experience. It is also quite (nearly 80 percent) likely that those Liking Hello Kitty on Facebook are Democrats, but even more (about 90 percent) are those that Like rapper Nicki Minaj and *The Colbert Report*.

Kosinski and his group asserted in 2015 that based on our Likes, computer algorithms can arrive at our personas more dependably than friends, parents, and spouses can.[139] Computers beat us in chess long ago, but are machines already better than us at reading people?

On this point we can call a halt and say the computer only has five personality traits to work with – and that such a scorecard could never encapsulate our entirety. Metrics allow us to quantify ourselves with numerals – without a need for a honed ability to comprehend individuals. Data analysis can confuse: we stare down the soda straw of what can be measured, and abandon more important things that are more difficult to examine. Kosinski and his team write: "Because the Big Five personality traits only represent some aspects of human personality, human judgments might still be better at describing other traits that require subtle cognition or that are less evident in digital behavior."[140]

The ability to predict what a human will do using models based on Facebook data, however, is nothing to snore at. The authors write: "Marketing messages could be tailored to users' personalities; recruiters could better match candidates with jobs based on their personality; products and services could adjust their behavior to best match their users' characters and changing moods; and scientists could collect personality data without burdening participants with lengthy questionnaires."[141]

In data, they warned, there are also dangers. People should be in possession of the right to control their own digital footprint. Lawmakers, regulators, and service providers should take steps to ensure private information stays out of the wrong hands. Personality information can offer an attractive tool to those who are just out to use us, they cautioned.

Pandora's box, though, was already wide open.

Where There Is Emotion

In 2014, Kosinski's colleague, Aleksandr Kogan, took an interest in the Facebook data that Kosinski had gathered. He wanted to monetize it, and asked permission for access, but Kosinski refused. Kogan then elected to collect his own base of essentially the same data – just much more of it. As his Trojan Horse, Kogan deployed a personality test on Facebook entitled This Is Your Digital Life. About 270,000 people completed the test and agreed to divulge their profile data – without knowing, of course, where it would end up. An app developed by Kogan absorbed the profile information of also the nearly 87 million Friends of the test takers. This is how Cambridge Analytica, the firm that helped Trump win in 2016, got the keys to sway America.[142]

A former Cambridge Analytica employee, Brittany Kaiser, said in spring 2018 that Kogan was not the only one gathering such data. The number of people duped into fueling the phishing spree is therefore probably much higher. In addition to everyone's digital footprint, Cambridge Analytica had an immense catch of other info at their disposal – such as the prior political activity and consumption habits of voters. The CEO of Cambridge Analytica, Alexander Nix, touted about how they had accumulated 4000-5000 datapoints on every adult American. The more that is known about an individual's behavior and preferences, the better the chances are that they can be played.[143]

There was a powerful belief within both the Trump campaign and Cambridge Analytica that intelligent ideas and facts are less important than the hopes and fears of the people. "The big mistake political parties make is that they attempt to win the argument rather than locating the emotional center of the issue, the concern, and speaking directly to that," said Mark Turnbull in an undercover interview with Britain's Channel 4 News. "Our job is to drop the bucket further down the well than anybody else, to understand what are those really deep-seated underlying fears, concerns."[144]

Political advertising used to serve as a source of alternatives that voters *selected*. Emotions and impressions have always mattered, but targeted ads guided by data infiltrate the subconscious in a far more cunning way. The recipe goes something like this: first, you relinquish some personal info, which is used to conduct a personality and behavioral analysis about you, on which basis you are then prodded for weak points with things that you are known to grow gooey over. Then you are stroked into the right mood with a targeted message or a piece of fake news – something that speaks to you, despite possibly seeming unfeasible. And with that you start moving away from your favorite toward someone else or give up on voting altogether.

Former Cambridge Analytica contractor and whistleblower Christopher Wylie has said that the firm customized targeted messages and videos to conform to personality type. The strings of the agreeable and extroverted are pulled in a different way than those of the neurotic introvert. The latter are browbeaten with fear, while the former get the message in a more joyous manner. Wylie told NBC that: "The premise is not necessarily to do standard political messaging where people are aware that I am trying to be convinced of something. It is to change what people think and perceive about what is real."[145]

The degree to which these methods influenced election results will become clearer at some stage in the future. Neither Trump's campaign nor Cambridge Analytica had anything close to the world's ideal brainwashing machine. Many researchers remind us that the capacity of personality analysis to shift opinions is limited, but the elections were decided in three swing states by a voting margin of 80,000, making minute amounts of competitive advantage irrefutably meaningful. Marketing research suggests personality-based ads get more clicks and cause purchases to go up by tens of percent.[146]

At any rate, we ought not to chalk up Cambridge Analytica, which was bankrupt by May 2018, as just an exception in which some slimeballs in made-to-measure suits violated privacy and jeopardized democracy. The firm might have dropped the bucket far down the well, but many others have done the same.

Signals were already visible in Barack Obama's 2008 and 2012 campaigns. Zeynep Tufekci wrote in *The New York Times* in autumn 2012 about how the utilization of big data in targeted ads threatens to change politics completely. In previous decades, according to Tufekci, influencing voters was at least relatively transparent. Everyone was equally privy to the same public message and incorporated it into their vote. The bashing of opponents on television or in the press is a side effect of any open society, but it was once possible to actually present a counterargument to allegations.

The opposing party and the media are not provisioned to fight against individually targeted advertisements that go unseen by the rest of the general populace. Targeted ads also enable campaigns to be two- (if not multi-) faced. Not even by mistake would a voter that visits his church's page be treated to messages in which the party says they support gay marriage. It is not disclosed to those subsisting on welfare that the party is pushing for tax breaks and cuts to social services. People are rather told something that their internet behavioral traits indicate they want to hear.[147]

Obama made revolutionary use of data during the 2012 presidential election, and managed to mobilize many of those to whom his opponent, Mitt Romney, was oblivious. Obama's data scientists, like Cambridge Analytica, snatched up also the profiles of his supporters' Facebook Friends without consent. The campaign's former director of data integration, Carol Davidsen, has said that: "We would ask permission to basically scrape your profile, and also scrape your friends, basically anything that was available to scrape."[148]

During the 2016 election in which Trump was the victor, data professionals from analytics firm BlueLabs assisted Hillary Clinton with tailoring her messages. The UK campaign demanding Britain's exit from the EU also used data to influence opinion, and there is evidence that the opposing campaign did the same.[149]

Some might be armed with better algorithms than others, but new data-based methods of phishing for voters look like they are here to stay.

Dangerous Giants

In the backlash against Cambridge Analytica, Facebook took a blow to its own prestige when word came out that the company had (without user consent) opened profile information not only to Kogan, but several other app developers too. Facebook also disclosed user data to parties such as mobile phone makers.[150]

We should ask, though, is Facebook such a troublemaker simply because it put information into the wrong hands? Not in the least. Let us take note of the fact that Facebook's true customers are not us users, but rather heavyweights in the commercial and political world. Its business model floats on the buoyancy of spying on users and the transmission of ads crocheted together from a broad database of their uploaded information. Facebook has also accumulated data about what people do when not prancing through its pages, or even online; for example, it has acquired from data vendors information about consumer buying behavior at brick-and-mortar stores. Facebook enjoys at least the same leverage to influence moods and views as Cambridge Analytica.[151]

And it has tested it. Facebook collaborated with Cornell University on a study in 2012 where, unbeknownst to those receiving them, some users were shown more negatively oriented updates than usual while the rest of the participants were shown more positive. And what do you know – fingering around with the newsfeed turned the users' status the very emotional tone that they were exposed to. The effect was not earthshattering, but it proved to be an influential factor nonetheless. An article summarizing the study came out in 2014, leaving many agape at Facebook's ability to move people.[152]

Less attention was paid, however, to the experiment Facebook conducted in cooperation with the research group of Robert M. Bond. During the 2010 US congressional elections, Facebook sent voting reminders to over 60 million users. Those to whom it was conveyed with photos which of their Friends had reported having already voted eventually upheld their own solemn civil duty 0.39 percent better than average. The operation also yielded collateral effects – the Friends of those receiving reminders exhibited a stronger than average voter participation rate. The researchers estimate that their intervention resulted in 340,000 more votes making it to the ballot box.[153]

An increase in voter turnout is good, and the participants in this study were selected at random, but if Facebook picked those they cajole into voting, it would have had an exorbitantly influential hand in the election's outcome without anyone knowing. Some Facebook employees have publicly scorned the company practice of suppressing conservative news.[154]

Facebook has also celebrated itself on its own website for being able to "influence voters." In the 2016 presidential election, it helped Donald Trump, Hillary Clinton, and Bernie Sanders to target their ads, no doubt doing it by employing the personal information of its users. Some are of the opinion that the Trump campaign's "Project Alamo" proved more pivotal than Cambridge Analytica. With Facebook's micro-targeting tools, it was able to reach and speak effectively to the target group. Trump campaign workers, for example, sent black potential Clinton voters in certain states animations of Clinton referring to African-American youth as "superpredators".[155]

Enjoying perhaps even more potential than Facebook to influence people is search engine Google, whose primacy descends from the unpassable ease that it brings to the modern info search. Should Google wish to pick who wins elections, it could pull it off without much effort or anyone knowing, says the work of psychologist Robert Epstein. The first two results of a Google search are endowed with a disproportionate amount of power: they are often accepted as absolute truth. Epstein's research shows how the manipulation of the initial search results can change what people think of politicians. Support for a candidate can go up by more than 20 percent among those who have yet to decide who they will vote for (while across the whole electorate the increase is usually no more than a few percent). Regulators might find it challenging to determine which results came about by tampering, because Google fandangles its data into personalized search results that conform to the user.[156]

The notion that search results are manipulable is not a figment of our imagination. According to the EU, Google has already unfairly favored in its search results its own Google Shopping price comparison service. The EU commission slapped the company with a 2.4 billion Euro fine for that in 2017.[157]. Although Google – or any nationally dominant search firm – might not knowingly have its hand in the results, search engine algorithms amplify shifts in opinion in ways that cannot be predicted.[158]

The Vacuum Trade in the Internet Age

We live in a surveillance economy. Businesses survive or die on the basis of whether or not they know their customers, and that is why they always recycle the same bit about how consumer spying (both on and off the web) is in everyone's interest: it is thanks to snooping that we can sun ourselves under ads for crap that lathers our fancy. Should you have a soft spot for anti-technology literature, Amazon will provide a list of books for your perusal.

If you rather prefer expensive shoes, images of them will keep finding their way onto your screen. Many consumers enjoy this (things often transpire in such a way that buying a ladder, for instance, convinces the algorithm that we want a hundred more).

Targeted advertising makes us monocular. It limits the diversity of our world, and it takes alternatives off the table. We are shown only those things in which we have been observed to be interested – things that people *like us* are known to get excited about. Targeted ads can turn us into our own caricatures.

Occasionally companies demonstrate that they know a little too much about us. That is certainly what can be gleaned from the high school girl in America that began receiving coupons in the mail from Target for baby clothes and cribs before even telling her parents that she was pregnant. Target's algorithms had interpreted the girl's weeks of prior purchases as those typically consistent with someone expecting.[159]

The pinpoint placement of ads offers many companies a shortcut to fortune. Those shoving products down the throats of net shoppers are today's version of the door-to-door vacuum salesman. Having data available about potential customer weak points makes the process of plying for buyers easier than it was for the traveling sales rep. If data suggests a man is concerned about his sex life, he is carpet bombed with Viagra ads. If someone is seen as being in financial duress, they might be offered consumer credit, but at ungodly interest rates.[160]

The more heinous cases of ads being guided by data are found in America, where those in the lower class fighting for social status are hawked by scamming and overpriced schoolhouses. The for-profit career training institute, Vatterott College, stipulated internally who their targeted ads should aim for: "Welfare Mom w/kids. Pregnant Ladies. Recent Divorce. Low Self-Esteem. Low Income Jobs. Experienced a Recent

Death. Physically/Mentally Abused. Recent Incarceration. Drug Rehabilitation. Dead-End Jobs – No Future."[161]

Sorry, But You Were Not Selected by The Machine This Time

The problem does not just lie in the fact that personal information is used in the manipulation of people. Big data today is where we go for antidotes to a broad swath of issues once addressed by humans with an ability to use their own judgment. Data's take on our past enables machines to forecast our future. Can we repay the loan, are we a good fit for the role, do we pose a risk of committing a crime later? To all of these, machines profess to have the answer.

Monetary lending decisions in the future will be automated. The software will give a thumbs up or a thumbs down based on the algorithm's calculation of how likely the loan is to be repaid. Machines are already often in charge of evaluating unsecured loan applications at many banks. Algorithms are also usually allowed to determine who gets approved for short-term cash "quickie" loans.

Sometimes the machines pump out decisions that are preposterous. In spring 2018, The National Non-Discrimination and Equality Tribunal asserted that the algorithm used by payment solutions provider Svea Ekonomi discriminated against a Finnish man who was seeking a loan. In assessing his credibility, the algorithm took note of things such as gender, age, place of residence, and mother tongue. The firm's accumulated data evidently found young Finnish-speaking males from Dispersed Settlements too risky, and thus rejected him.[162]

The CEO of credit scorer EFL, Jared Miller, told *The Economist* that in the future, social media and other internet history may serve as better indicators of an applicant's ability to service debt. Multiple banks have implemented EFL's personality test program in their loan applicant assessments.[163]

Data-devouring algorithms enjoy an increasingly important role in the recruiting process in the United States. Large chunks of time no longer need to be poured into applicant analysis when the algorithms sift through the pile and pull out the closest matches. Recruiting schemes based on big data also rely on the digital footprint of the applicant, and with this in hand, the machines can quickly rustle up an answer to how closely an applicant aligns with what *usually* equates to a good worker.[164]

What if the shortlist landing in front of the manager calls applicant X less than bright for liking Harley-Davidson and Lady Antebellum on Facebook – despite he or she having the intelligence of a Nobel laureate? The hiring manager is typically unaware of the basis on which the machine picked candidates – he just accepts the results of what the algorithm in its infinite wisdom spits out from stats it was able to correlate and report like some angel of the Oracle. The algorithms also do not necessarily learn from past mistakes. If, for instance, they manage to discard a strong applicant that later becomes an expert in their field, the machine won't see it, and will just bumble along on the same binary balderdash. Big data in America is harnessed also in measuring employee creativity and determining who gets the boot. Decisions made by algorithms are nearly impossible to appeal, due to them being protected under the opaque veil of corporate property.[165]

We can ask this of algorithms in recruitment: what if it is observed that a woman can fulfill a particular duty much better than a man – would it then be right for the machines to fill the shortlist with only women? What about the opposite case? Algorithms are liable to just reinforce existing gender monopolies in certain professions, as well as perpetuate the practice of treating people as dimensionless informatics – and not as individuals.

Some also believe automated recruiting decisions could be used to more effectively pair applicants with jobs that suit them.

From the standpoint of individual dignity, however, is it reasonable to think that an algorithm could equip everyone with the perfect career path? Is the solitary and dogged search for self – the learning of things through failure – not integral to the spice of life and making something of oneself? The previously mentioned group led by Michal Kosinski believes that "in the future, people might abandon their own psychological judgments and rely on computers when making important life decisions, such as choosing activities, career paths, or even romantic partners." (The Transhumanist Party's former US presidential candidate, Zoltan Istvan, revealed during my 2015 interview with him that he believes that within the span of a couple of decades, algorithms will be making political decisions for us.)[166]

Future generations will therefore no longer embrace the esprit of Sinatra's *My Way*, and will struggle to find substance in the suggestion that we are richer with our imperfections and failures than without them. In the future, we will perhaps join together to sing a newer, more progressive chorus: *Not one mistake have I made in life, since I spent it buried in my smart device...*

You Are Guilty of a Crime That Hasn't Been Committed

A Helsinki University study finds that the vast majority of us have at one time dreamed of killing someone. Appalling thoughts are part of our nature – and Googling about them is private. There is evidence, however, that searching for information about crime and suicide on Google is connected with the commission of real crime and suicide.[167] Is there probable cause then for officials to comb through web searches pertaining to murder in the name of public safety? Would it be appropriate for police to kick down doors just because of what someone Googled?

Philip K. Dick's sci-fi novel *The Minority Report* – like the film starring Tom Cruise – presents a world where people are taken into custody before they even cross the law. In Dick's narrative, this is enabled by clairvoyants who bequeath their lives to the prevention of crime. The operations seen in *Minority Report* will soon be possible thanks to the development of big data.

In the summer of 2013, on the infamous West side of Chicago, the police came and knocked on the door of Robert McDaniel. He hadn't ever been charged with a violent crime, but the cops cautioned him to behave, saying they had their eye on him. The 22-year-old high school dropout had managed to make it onto a list of 400 people dug up from data placing him as someone likely to commit a violent crime in the near future.[168]

This data was not assembled from Google searches, but rather a mass of background information put together by police and fed into machines. McDaniel's presence on the list was more a consequence of his shady circle of friends and one minor marijuana-related infraction. The algorithm had extruded a match with those of similar backgrounds that had either committed or suffered from gravely violent crimes in the past.

Police in European countries are not usually armed with the same kind of systems as Chicago officials, but even in Europe the cops can show up at your door to discuss your internet use in the name of safety. Finnish National Board of Investigation Detective Superintendent Sari Sarani says this might be done to those who make noise on the net about causing mass destruction. The police could also pay visits to those who have downloaded material to their computers from the anonymous Tor network about how to abuse children without getting caught (the police obtain info about user identity for these from Interpol or Europol).[169]

Sarani says it would be nice if they could use big data in the same way as New York City, where the drop in violent crime can be partly explained by the CompStat system, which assists police in focusing their resources more effectively. CompStat uses statistical analysis to communicate when and in which area crime will likely occur. With this information, police manpower can be redistributed to places where it is needed most. Finnish law does not (yet) allow the use of data in this way.[170]

In the United States, big data algorithms are exploited more in assigning threat scores to suspects and convicts. In some states, the machines are used in the administration of actual verdicts, while in others only when evaluating the possibility of parole or setting bail. Several elements that have nothing to do with the crime can affect the length of time someone gets incarcerated. What kind of family did you grow up in? Have any of your family members or friends been charged with a crime? What kind of education do you have? Where have you lived? Are you unemployed? Etc. Etc. And while the factors fed into the algorithm are clear, little is known about how the machine arrives at its pointed conclusions, nor is it easy to present a case against them.[171]

This breed of individual-focused big data utilization can incriminate people for just being the person they are and having their own history – rather than for any crime they may have committed. It is a fact that in the United States, blacks commit more crime than whites as a percentage of the population. Homicides are approximately seven times more likely to be committed by blacks than whites, but is it then right to extend blacks a more stringent sentence for identical acts, or make it harder for them to get released on parole? Never. Every individual should be treated equally, regardless of ethnicity, but glaring disparities still persist. A study found that predictive algorithms used in Broward County, FL put incarcerated blacks as far more likely to repeat their offenses than whites. But only 23 percent of whites examined stayed out of jail within 2 years of their first arrest, while the figure for blacks was 45 percent.[172]

Delegating decision-making to machinery is rooted in the assumption that they are perfectly objective, but it legitimizes discrimination on the basis of ethnicity. The issue, though, is not just about race. As a matter of due process, the unignorable irrationality in the room is that being born into a broken family, living in an iffy area, being male, or being poor might put you in jail longer for no other reason than that the algorithm found you statistically more likely to commit a repeat offense.[173]

It is without a doubt desirable that we try to reduce the commission of health and life-threatening crimes with new technology, but measures to minimize wrongdoing cannot be implemented by just any means. It is a different thing to surveil high crime areas than harass innocents or convict people with flimflam from a database. There is no logical endpoint to enhancing the capacity of our handlers to tighten the leash.

Making a List, and Checking It Twice

New technologies have supplied governments with an increasingly more opulent buffet of ways to spy on their citizens. American intelligence officials are able to turn almost any home electronic device into an undercover agent. New intelligence legislation in Finland puts broader powers in the hands of the government to do whatever is technologically possible. Pervasive electronic spying comes dressed in the pretense of making us safer, despite a lack of evidence supporting its utility in preventing terrorism.[174]

Electronic monitoring has also empowered authoritarian regimes to put more muscle behind gathering information from their nation. One important technology making state spying and control easier is facial recognition, with which the face of anyone can be connected with their correct identity. AI can unearth from security camera footage someone wanted for crimes – a task Chinese police often accomplish with smart-glasses. Facial recognition, however, makes it easier than ever to keep dissenters obedient, silent, and scared.[175]

The technology is being taken still further. Many found

Italian physician Cesare Lombroso laughable for theorizing that criminal acts have such profound genetic origins that those committing them can be identified by their facial features and cranial shape. Is Lombroso so totally cuckoo considering how machines now pick criminals out of crowds with a 90 percent success rate using *only facial features*?[176]

VGG facial recognition software can distinguish homosexuals from heterosexuals with similar certainty. With only one photograph for reference, the software has an 81 percent success rate, but with five, that jumps to 91. Lesbianism to this system is about 10 percent less visible than homosexuality in men[177], but what really starts ringing the alarm bell is the fact that being gay in many countries is a crime – one that some punish with death.

Leaps in facial recognition technology do not present us with a conundrum just because some authoritarian states are in a position to abuse it. Progress in the field can totally poison the privacy that a purposeful life requires. The software is already capable of taking respectably accurate readings of a person's emotional disposition, as well as determining if someone is lying.[178] In school, we were all taught that lying is bad, but without white lies, the world would be a more distressing place. The violating of emotional privacy with new technology should not be permitted, but if it becomes normalized, I would suggest that everyone get a balaclava.

The digital surveillance society has perhaps been taken further in China than any place else, to the degree that schools keep pupils alert and lively now with facial recognition tech.[179] Cameras do not stop recording when teachers turn their back on the class. China's choice to start assigning behavior scores to citizens dropped jaws overseas. Ratings rely to an extent on the participation of the individual, and because paying for things with your phone is widespread in China, transactions leave a trail of data about everyone's movements. This information is coalesced into a registry from which each person receives their score. In addition to your impetuousness for buying, scores are dictated by level of education, as well as how well one handles one's pocketbook and debt. It behooves one to pick one's friends cautiously in China, because being connected with the wrong people on social networks can lower one's score. This "social credit rating" that seems eerily similar to the *Black Mirror* television series is up and running among us today.[180]

Should the machine assign you a high score for behavior, you could stroll past the line for security checks at the airport. A low score could prohibit you from traveling by air and leave you unable to get a loan. It is envisioned that one's social and political habits, both on the net as well as off, will weigh heavily on received ratings. The point is of course to press-fit people into the receptacle of model citizen.[181] They are making a list and checking it twice to see who's been naughty and who's been nice!

But being surveilled everywhere will leave scars. Social theorist Jeremy Bentham's obsession in the late 1700s was the Panopticon, which, according to his design, advised that penitentiaries, for example, should be constructed so that inmates are unaware of when they are being watched. That way, they learn to conduct themselves under the assumption that they are being constantly monitored. Bentham believed that over time the inmates would internalize this omniscient eye and alter their behavior according to it.

The double agent in our homes rewires our heads and affects what we dare to say and think – regardless of whether we are on or off the net.[182] It becomes therefore safest to always behave as prescribed.

You Have Committed a Thought Crime

Our world is not the same as the inconsolable dystopia of George Orwell's *1984*, where people are granted just as much hope as they are granted a boot "stamping on a human face – forever". [183] The loss of privacy in our society does not manifest itself as overt misery, which is why few of us find it perishable, let alone lose sleep over it. No one forces us to Google stuff, Like things on Facebook, or go porn surfing. We infect ourselves quite voluntarily, and give every home's double agent our info of our own volition.

Not only do we cough up our data without being told to, but any person in our vicinity with a smartphone can take photographs of us and then share pics, vids, and other scuttlebutt with anyone. There is nothing for big brother to do when every eager beaver is both informant and jury. Author Walter Kirn has said of this little brother surveillance that "The invasion of privacy...has been democratized."[184]

At the same time, the foothold that this technology has gained in every corner of society has made a mess of our mental privacy settings. Many of us can no longer delineate between what is and what is not everybody's business. It has somehow become the new normal that people receive public lashings for private shenanigans.

Charges over the commission of thought crime are often entered without any sense of proportionality, as indeed was the case when newly admitted students to Harvard posted some politically incorrect jokes on Facebook during the 2016-2017 change of semesters. After institution officials got wind of it, they rescinded the admission of at least ten prospective students to whom it had been extended.

Perhaps the most abhorrent example, however, is that of Jewish billionaire Donald Sterling, whose black girlfriend, V. Stiviano, recorded the racist slurs he vented during a 2014 spat on her phone. After it went public, the NBA forced Sterling to sell his ownership of the Los Angeles Clippers, slapped him with a 2.5 million dollar fine, and banned him from NBA participation (even as a spectator) for life. The NBA divvies out unbelievably more lenient punishments to its players for using prohibited substances or resorting to violence. Darren Collison, for instance, was just benched for eight games in October of 2016 for battering his wife – which isn't a private matter.[185]

When we shake the public finger at private outburst, we lower the bar to petty shaming. Those banging the gavel in these cases can soon find that they themselves are now the target, because rarely do any of us walk unencumbered by past comments that could give off a stench in the right sunlight. Seventeenth-century French king Louis XIII's chief minister, Cardinal Richelieu, is purported to have said: "If you give me six lines written by the hand of the most honest of men, I will find something in them which will hang him."[186] In the internet age, there are millions of phrases through which both big and little brother are diligently combing. All data (even private) can be used against us – so eventually our freedom to think is in danger.

Curbing the Data Giants

Privacy is becoming a scarce commodity, but each of us in our own way can regulate the extent to which it is violated. The Tor browser, for example, anonymizes internet activity and blocks third parties from tracing your data. The encryption that accompanies Tor will of course slow throughput considerably. The TrackMeNot extension, instead of cloaking internet traffic, drowns it in background noise. Your real search gets buried in the mayhem that the application generates. If there are no methods to pull real searches out of the static, it is virtually impossible to obtain anything credible from them.[187]

Reigning in the data collection hegemony of Google and Facebook would be a tall order, but in the end, we simply do not have to use them. There are functional alternatives to Google, like the privacy-protective DuckDuckGo, and there has long been an alternative app to Facebook called face-to-face communication. Tracking and tracing a smartphone is far less strenuous than a computer, but here again, we are free to just choose not to use them. I myself have gotten by without a smartphone my whole life.

Cyber security firm F-Secure's data security chief, Erka Koivunen, is hardly surprised that people do not have the time, interest, or will to use privacy-honoring apps at every turn. "The most important area in need of change is for lawmakers to bring the tech giants under control."[188] As in earlier periods of rapid technological and economic advancement, injustices inherent to any new paradigm can be offset with regulation. Governments ought to get on this – and start codifying law that serves the people.

There has traditionally been a marked difference between the EU and US approaches to privacy – something that is also visible in their respective statutes. This difference reached new heights in May 2018 with the activation of the General Data Protection Regulation (GDPR), according to which net users shall give their approval before any service provider gathers their data or profiles them. If, however, the user does not agree to this, they forfeit the right to use said services, making the selection of available alternatives very limited. The GDPR also requires companies collecting personal information to inform their customers of what is being collected. The regulation applies to all EU citizens – including those not living in it.

Erka Koivunen does not see the GDPR as threatening the private data collection model of internet business, but the ePrivacy Directive might do that, if you ask tech lobbyists. The ePrivacy Directive might enable the use of internet services without having to consent to data collection.[189]

Of primary importance to us is enhancing the transparency of the algorithms that wield hidden hands in our lives. Data scientist Cathy O'Neil insists algorithms should be displayed for public scrutiny and gauged for fairness. O'Neil's belief is that algorithm-led decision-making should be discontinued in those areas of life where they do not work well – one example being the rating and firing of teachers in American schools. New EU data protection regulations highlight the right of individuals in certain instances to receive information about how an algorithm made a particular decision (Article 22). Many legal experts, however, interpret this to mean that companies are not bound by anything.[190]

The use of big data in politics is also screaming for transparency and constraint. Sociologist Zeynep Tufekci has proposed that election campaigns be forced to disclose the targeted messages they send in a public database – an obligation not much different than the reporting of campaign financing.[191]

Alphas and Epsilons

The more information is fished from us and the more subserviently we give it, the more relevant big data becomes – and when information becomes more relevant, it can be more effectively applied to predicting the future. Big data is being sucked up with shameless arrogance today from a broader swath of areas in a bubble of ambivalence about where it will ultimately be used. An unquenchable thirst for data often fosters a desire to do good, but as we know, the road to hell is paved with good intentions.

One example of especially virtuous intentions is neuroscientist Katri Saarikivi's NEMO project, whose purpose is to gather emotional data from net users with which to expand empathy in cyber encounters. In what situations while net surfing does heartrate quicken, when do the pupils dilate, when do we experience cold sweats and despair? The idea is that the information indicated by these kinds of emotions is transmitted in observable form to our online correspondents. Our perceptions toward whomever we are conversing with might be more compassionate if we can detect when they feel bad.[192] It feels equally bad to think that this very information could in the end just be used against us. Emotional data is a chronicle of our deepest and most private reactions, and it would best if it was not meticulously sifted through.

The near future promises an internet that hosts more voluminous and precise information about our genes and health condition. Many surrender their genetic data because they believe it will assist in the treatment and prevention of sickness, but this health line is always a hawsepipe for a long list of less presentable types to stowaway on (those donating their genetic data to research should know that they will likely not remain anonymous for very long, because computer software today can already generate a workable facial image from one's genes. Progress in this technology will enable facial recognition programs to pair faces with their exact identities).[193]

This information is also sought after by the likes of former Cambridge Analytica employee Aleksandr Kogan.

The psychologist tried negotiating with researchers at the Harvard School of Medicine for access to their store of health and genetic data on Americans. "Can you imagine the possibilities then?" Kogan wrote to his Cambridge Analytica colleagues in February 2014, adding "It's going to be AMAZING." Kogan failed to obtain the data, but Google-owned DeepMind has an immense bank of British patient health information on hand for research. Cyber hacks have ensured that similar health information is copiously available for sale on the so-called dark web.[194]

In the near future, data about us will begin to be collected by smart home devices – you know, the ones that run our smart houses, to which we drive in our smart cars. How in the world have we gotten by until now without refrigerators that connect to the net?

Earlier we discussed how accurately one's personality can be determined simply by analyzing Facebook Likes. Today, internet footprint and buying behavior are being used to add to the pot our emotional portfolio, our genes, our lifestyle, and our health. From this mixture, we ought to get an even sharper glimpse of our true essence and intentions. Big data provides a more precise assessment of our persona, our predisposition to disease, and soon our aspirations, our future spouse, and the professions of our kids (scientists believe big data will allow us to forecast evolution – but, for the time being, only for very simple organisms).[195] All of these predictions are established on the same utilitarian logic with which we are familiar – that it will allow us to live a bit longer, be a bit healthier, and not have to endure so many unpleasant surprises .

But does life not involve uncertainty? Or is it better to just wait for whatever disease lurks in our genes to self-actualize? Would that not be comparable to watching a hockey match whose final result is already known? Is the marginal increase in average lifespan so valuable that it can come at the expense of the death of adventure and surprise? Is achievement not founded on the fact that applying ourselves to something does not include a guarantee that we will succeed?

The notion that everyone shall be extended equal opportunity in life might also gradually begin to be questioned, because it descends in part from the belief that there is a lack of credible information about human potential and intent. Erka Koivunen asks: "What if with big data we could dependably foretell that such and such will always be a pathological liar, such and such will climb to the top of the social ladder, such and such will never grasp abstract concepts, and such and such should immediately be thrown in jail?"[196]

Many have perhaps caught on to the fact that dreaming can often lead to disappointment. It is more meaningful, however, to dream and be disappointed than be fed scientific findings from the start saying your dreams will not come true. We do not need to know who was born an alpha, and who an epsilon.

5. The Death of Purpose

How our spiritual and social lives are suffering from our obsession with technology

Alas, the time is coming when man will no longer shoot the arrow of his longing beyond man, and the string of his bow will have forgotten how to whir!

I say unto you: one must still have chaos in oneself to be able to give birth to a dancing star. I say unto you: you still have chaos in yourselves.

Alas, the time is coming when man will no longer give birth to a star.

Alas, the time of the most despicable man is coming, he that is no longer able to despise himself.

Behold, I show you the last man.

'What is love? What is creation? What is longing? What is a star?' thus asks the last man, and he blinks.

Friedrich Nietzsche[197]

For what will profit a man if he gains the whole world and loses his soul?

From the Gospel of Matthew, verse 16:26

German writer Johann Wolfgang von Goethe (1749-1832) pondered this bit from Matthew with Heinrich Faust, the main character in his play. Faust was ready to sell his soul to the devil in exchange for the fulfillment of his worldly pleasures. Goethe knew that the modern human's quest for a better tomorrow is limitless. As a society, we have gobbled up progress without regard for what it costs, and the tragic piece of it is that improved living conditions do not usually improve our spiritual space. We are forced to confront life's underlying trials whether we hail from the apex of an advanced technological civilization or not.

As Mephistopheles (the devil himself) said of man in Goethe's Faust:

> For he demands from Heaven it's fairest star
> And from the earth the highest joy it bears.
> Yet all things far, and all things near, are vain,
> To calm the deep emotions of his breast.[198]

We have achieved materially much more than those of preceding centuries could have dreamed. Still (or perhaps precisely because of this), man grapples for purpose more than ever before. While the whiz kids of Silicon Valley see their new devices and apps as *pulling* humanity out of a pinch, in many cases they are just pushing us into them. When technological advancement gives us something, it takes something else away. What we receive is usually easily quantified, but what we give up is usually valuable beyond reach.

In this chapter I examine the intangible price of progress – the spiritual[199] and social consequences that come from our compulsion to advance. The digi-obsession that consumes us creates a culture of insatiability and derails the social engine of purpose. The internet is inherently a place that favors those who fare well, and rebukes those who don't – leaving us with an irregular distribution of joy in social life and a myriad of mental illnesses.

Easy or Fun

The central goal of technological development is to make life easier, but its pursuit cheapens our experience. Purpose in life does not emerge from convenience.

Material progress has brought enjoyment-by-keystroke, which we – unlike the hunter-gatherer – can always spoil ourselves with here in the welfare state. Any of us can roll by the store for a bag of sweets or a six-pack of booze, then shove it all down in one sitting. We can also get on the net and beat the meat all day in a personal peep show before the world's newest moguls of porn. In the virtual world, we receive easy acceptance from our fleet of ostensible friends. Had Mephistopheles presented all this to Dr Faust, it would have ruined him in short order. Enjoyment-by-keystroke provides a wide array of momentary delight, but few find that it provides purpose.[200]

Man is an organism that is capable of being satisfied with his life only when he constructs it himself. We seek struggle to justify our joy. We need challenge to lead us to achievement. The setting and reaching of (not too) intimidating goals are key components of what gives us meaning.[201]

Enjoyment obtained through tribulation therefore lasts longer than what comes from pushing keys, and this is also why reading books brings us a deeper breed of thrill. It is why success at work is important to us, why we value the creation and maintenance of human bonds, and why we derive so much satisfaction from the building and remodeling of our own home. None of this is straightforward or self-evident – it takes guts, tenacity, discipline, and endurance. It is what makes it all worth it.

Some have tuned in to the fact that technology's job is not to do everything for us. We want to engage in tasks that we find fascinating, even though it might make life more difficult. Extreme sports like marathons and triathlons have strong followings, knitting up one's own mittens yields a feeling unobtainable by buying them at the store, and climbing a mountain to take in the view provides something altogether unattainable by riding the lift. Some of us still prefer to cook and clean by ourselves, although robots can now do both of these well.

At the same time, the modern human wonders if it is foolish

to not use machines for things that they do better. But we should forget this – because there is no point in courting the notion that beyond the maximization of convenience, efficiency, and speed there are no things of value. If that were true, we ought to just brand our intangible experience as irrational and throw it away.

A Culture of Insatiability

The penance for watching progress send ease, efficiency, and enjoyment into infinity is our own insatiableness.[202] The Westerner is used to abundance and unending alternatives, and he can no more settle for less than he can stop reaching for inebriant. As Pauli Hanhiniemi sings, "that which yesterday turned me on, today just makes me yawn."

The internet offers nearly everything – a one-stop shop of the world's grandest art, its funniest stand-up comedy clips, and access to the music and people that we seek most at that moment. Such plethora is more curse than blessing. The principle of financial inflation can also be applied to questions of culture: art, people, and entertainment decrease in value when in surplus. If everything is readily available, the purpose of everything blurs. It is impossible to yearn for something without which the thing itself would not be needed, but if we do not yearn for things, how will we experience fulfillment?

Popular radio and television programs before the internet's omnipotence were precious because they were served in installments. Losing track of time meant missing out on the program, or it mandated the VHS be set to record your *Ally McBeal*. Today, with the net flooded with interesting material set on permanent play or paused for later, even our favorites can succumb quickly to disinterest. In a marketplace of abundance, what once was rare and irreplaceable becomes a stockpile of rewinds – a puddle of skin deep and fleeting into which we do not care to wade further.

It is no longer easy to give everyone and everything the

focus, the time, and the consideration that they deserve – which is a tragic turn of events for both those who create and those who consume. A Facebook friend of mine recently regretted how there is nothing good to watch on Netflix. She started ten films, then cut them all short for being unworthy of finishing. After too much good, nothing can satisfy our wretched souls.

The to-do list of today's human includes scrolling through videos, pictures, and text messages on Facebook, Instagram, Twitter, Whatsapp, and some others that I am not familiar with. While riding the city bus, I often get a close-up view of the blistering speed at which young people flip through each of these feeds. I am gripped with an Uncle-like interest in what this stream means to the desensitized mind behind their glazed eyes. Sociologist Georg Simmel wrote at the beginning of the twentieth century about how the metropolis was making us more indifferent, because reacting to everything with emotion will eventually rob us of our sanity.[203] The same can be said of social media. And today – in places where there is broadband – this brand of big city detachment has already gone rural.

Little by little this culture of indifference and disposability is starting to affect how we engage with other people. Social media and dating apps make it possible for a fortunate few to jump seamlessly from one relationship to another. If someone new can be found with a single swipe, few will ever view anyone as undumpable. The brain runs along on the delirium that excitement awaits us around every corner. But love dies there, and we are left always lusting for something better.

This trend was already developing before the internet was even around. The late sociologist and futurist Alvin Toffler wrote about how the requirements of working life and advancements in transportation made frequent moving common over the last century. This loosened the bonds that existed between us, and it turned everyone we met into at best an acquaintance.[204] Social media intensifies this trend by broadly expanding our ability to jump between people. This makes lasting friendships and successful partnerships that much harder.

If material progress and the internet environment produce a culture of insatiability, then the economy needs it. Karl Marx wrote in the 1800s that capitalism's success is based on its ability to continually create new demand. Selling new knickknacks to someone first requires that they grow tired of what they have. Meaning, being satisfied holds everyone back, and today we laugh at the sensitive, the inexperienced, and the innocent for settling for what they have. The more people there are that retain their spiritual virginity, the less the economy grows.

Virtually Human

The digital transformation has redefined the social experience. Shooting two lines off to a friend in Messenger every other hour does not require us to put our persona on the line like a phone call or face-to-face meeting. If we get what we need from influential Twitter figures then we do not have to listen to the half-baked political views of family or colleagues. We have all been at the dinner table when everyone there is just fingering their own phone.

This is understandable. We want to get by with less effort. We try not to give too much of ourselves while getting as much as we can from everyone else. In other words, technology makes it easier than ever for us to be pricks.

While social media offers more channels for easy and irreproachable communication, it stifles the intensity and depth of our connections. Conversations become superficial flings. Digital contact provides "the illusion of companionship without the demands of friendship," writes MIT professor Sherry Turkle.[205]

As we migrate more toward interacting remotely, the consistency of our offline circle of social relationships becomes drastically thin. Research shows that phone use in front of someone with whom you are conversing wounds their self-esteem, as well as their sense of meaningfulness and control. Those enamored with their phones always part with a piece of life that will never be returned. One survey saw 24 percent of American social media users concede to having missed out on an important life moment while engrossed in making a social media post. The survey was conducted in 2012, meaning the percentage is now likely quite higher. It should be noted that important moments go unnoticed also by those flicking through their phones to make status updates to things unrelated to whatever they are doing at that moment.[206]

For decades, our cars have served as our private rooms – a womb-like haven in which we do not have to deal with people. Today, this antisocial sphere accompanies anyone with a smartphone and earbuds, and it is why it is no longer proper to talk to anyone at the bus stop or in the checkout line. Being spoken to was once taken to be friendly, but now such a thing is a violation of personal space. Those victimized by this have to remove an earpiece for the uninvited claptrap of someone they did not approve as a contact. Offline conversation in the real world is already regarded as harassment. In the age of the smartphone, chance encounters are gone, and the concept of community withers away with our grandparents.

While it is now established practice to flee from random interactions, rarely are we capable of even being alone with our own thoughts. Digital platforms always provide passive companionship, and reflection is also under threat. For the sake of inner balance, we need both a social sounding board and periods where we are not poisoned by anyone. Similar to how muscles grow at rest, the development of our personality requires its own time.

Exactly when machinery is starting to behave more like humans, we are starting to behave more like machines. We seek deliverance from the conventional mess of living. Chapter

3 discussed how the digital transformation dulls intellectual culture, but it does the same to emotion. Frequent use of mobile devices appears to weaken our capacity to detect expressions of emotion when face-to-face.[207] The unbounded ability of man to go anywhere and do anything dead ends in a dropdown menu, and the complicated nature of feelings is reduced to a few emojis and Likes. Should you be more than just your heartbeat, this will make you a reactionary flake.

But instead of it causing alarm, many find the dehumanization of communication to be a harbinger of good. Living humans take too much of our time and, quite frankly, are a pain in the ass. They exert and stir up emotion, impose constraints, and create frustration. Online communication makes it easier to be ideal, and on chat we can prune ourselves prettier than we ever could in person. The idea that digitalization improves communication has gathered strength particularly among youth. Our digital devices are also the place we go when we want to disengage from things – or at least certain people. A 2015 survey in America found that 47 percent of those aged 18-29 employ their phones from time to time just to avoid others.[208] While it might not be right to call these digiturds misanthropist, it would neither be inaccurate to call them species averse.

This wind shift in the way we communicate has begun to ripple into our appetite for empathy. Studies show a significant drop this millennium in at least American students' ability to put themselves in others' shoes.[209] The causal connections behind this are still up for debate, but we could call out digital devices at least as one suspect. Online chatter takes us away from each other. It is easier to forget, discard, and discredit someone who is not staring you in the face. The absence of nonverbal reactions makes learning to empathize more difficult.

The Futility of Being Genuine

When internet life becomes just an ideal, our regard for

authenticity fades. In her book *Alone Together*, Sherry Turkle recounts the time she visited a Charles Darwin exhibit at the American Museum of Natural History with her daughter, Rebecca. Giant tortoises from Darwin's revolutionary run through Galápagos had been installed as its drawcard, but Rebecca, unimpressed with their immobility, suggested they be replaced with robots that can move.[210]

Simulated life has begun to be superior to life itself. This course change is even visible in our definition of intimacy and our disposition to sex. Internet porn opens the door to sexual experience without the uncertainties of human relationships. There are suitable erotic playthings for all of our perversions, and finding the same thing in the physical world would be difficult. Lo and behold: research shows that the amount of sex we have has decreased significantly – exactly as masturbation has exploded. The role internet porn plays in the lives of young people has seen notable growth. Finsex statistics compiled by the Family Federation of Finland report a quarter of women aged 18-24 in 2007 watched porn in the preceding year. By 2015, that number was almost 60 percent. Young men consistently come in at well over 80 percent.[211] The interactions are not genuine, in-person, or reciprocal, but pictures and videos are sufficiently titillating.

Porn viewed under a Virtual Reality headset takes us one step closer to the real thing. Peter Rubin writes in *Wired* magazine that, "You're not watching a scene anymore; You're inhabiting it...It's far from pornographic and much more like human intimacy." A middle-aged VR porn user Rubin interviews says, "I remember the first time a girl whispered in my ear in VR – I could swear that I could feel her breath and the heat of her cheek radiating against mine. It sent tingles down my spine." For the most convincing simulation of authentic sex we turn also to sex robots, some of whom can be programmed with certain personality types, and others that are already equipped with a pulse. AI expert and author David Levy predicts that in the future we will love robots – even get married to them.[212]

One of the more well done works of recent cinema, Spike Jonze's *Her*, tells the story of Theodore, a resident of LA that falls in love with Samantha – the vivid voice of a computer program. The lovableness of Samantha (spoken by Scarlett Johansson) stems from her ability to know with digital precision the things that Theodore likes, and this is why he (and 8316 other simultaneous users) feel so pleased. Just like Cambridge Analytica, she knows enough about their preferences to pull the right strings.

Not everyone is necessarily as conservative and biased as I, but I believe love is not love if it comes as the automated output of pushing a button. It is an altogether different thing when after desperately searching, perspiring, blushing, and otherwise bumbling around we find someone whose journey merges with ours. When we keep giving machines permission to play pretend lover, it is simply too easy, too unspontaneous, and too mundane. In the words of the digigeek, this would be referred to as a computer game that we can beat without playing. It goes without saying that machines can no more be connected or love than they can despise or feel hate. There is nothing inside them – they feel nothing at all.

But still, we afford machines more than they deserve. We insist on wrapping what they say in the trappings of a human. We want to imagine that somewhere inside the transistor there is a soul. The degree to which we want to believe this astounded MIT researcher Joseph Weizenbaum in the 1960s during experiments on ELIZA, an early AI built to function as a Rogerian psychotherapist. Weizenbaum was appalled by how little it took for humans to conflate artificial with real. ELIZA was not so intelligent or empathetic as to be blankly accepted as human – belief in its humanness was just rooted more in subjects' desire for that to be true. It was "human enough" to convey that.[213]

Our trust in machines also chaperones us toward avoiding other people. When humans become lonely, machines fuel illusions about what others will do. A vicious circle forms where we self-medicate for estrangement, go virtual, then start actively seeking the absence of human contact. Human-ish then becomes human enough to meet the needs of those going it alone, and simulated emotion starts to become a sufficient substitute for many of our real feelings.[214]

Our Hearts Bleed For No One

If authenticity and meeting each other eye to eye are no longer distinguishable from their digital double, the danger is that the vast majority of us become mutually valueless. If the real thing enjoys no regard, why rummage around with regular people when you can go right to climax with the world's beauties in VR? Who wants to listen to the anecdotes of friends when Youtube is jam packed with the world's best comics?

The entrance of new technologies marks the exit of the chance encounter, and it is the sign that more and more of our social interaction is an orchestration. In theory, we are more capable than ever of selecting when and with whom we communicate, and such freedom of choice, at first glance, seems swell. The flip side, though, is found in the fine print written at the bottom of the user agreement: those who are not trending shall be banished to the fringe – alone and dispossessed. On the internet, that which garners the most interest immediately supernovas into a sensation, while everything else fizzles into irrelevance. Social media is a place where we can essentially deploy ourselves in duplicate so that the blast radius of our circle of influence is expanded. Simultaneously, fewer and fewer of us find ourselves desirable, fun, or intriguing.

There is no need to be tongue-in-cheek about drawing parallels between this and economic globalization. In the 1980s, there were still countless companies that, while not straddling the planet, got on fine – and whose products and services were happily procured for lack of anything better. But when getting by became a global free-for-all, those companies that were unprepared for it found they no longer had a place. Competition for the time and attention of friends at that time was similarly local. Being the funniest, smartest, or most beautiful person in the room was often the deciding factor, but these features are less gravitationally attractive in today's digital village. Social media has transplanted competition into the DNA of our personal relationships. When there are better alternatives to be had, no one has time for just someone around town. No one wants to settle for less, regardless of how mediocre one's company is.

This unpleasant development has only just started. Smartphones and social media have only been around for slightly more than a decade. People are equipped with enough of an appreciation for the prevailing way so that change cannot defeat the status quo in one swing. In the 1920s, sociologist William Ogburn wrote about the concept of *cultural lag,* a necessary step before the realities of technological change begin to alter the values, attitudes, and habits of people.[215] It is quite likely that after another decade those marred by a lack of retweetable trendiness will be shunned more harshly and just extradite themselves to a life of immersion in the virtual world.

The comment threads on social media make us acutely aware of our place in the social hierarchy, and in so doing diminish the psychological well-being of those that are not social. One big question that we will have to address in coming years is how to manage the mushrooming monstrosity of our sense of insufficiency. Will our virtual worlds be so ordered as to issue us both satisfaction and self-respect? We might be apportioned daily doses of porn and stand-up, but how many would count that as dignified life?

Two-and-a-half millennia ago, Plato put forth the theory that the human soul is composed of three parts: logos, eros, and thumos. Logos is reason, eros is desire, and by thumos Plato meant self-image – our efforts to get recognized. American political scientist Francis Fukuyama ranks this thumotic "struggle for recognition" as one of the more motive forces propelling humanity forward. It led to wars and territorial disputes, of course, but without it we would never have demanded democracy and equality.[216] Being human is obviously much more than just the mind's pursuit of reason and the body's drive to be satisfied. It is our mental vitality – the seemingly irrational fashion in which we vie for symbolic position. It is the thirst to be identified, approved, and empowered by others. Or at least it was...

The time is coming, Nietzsche believed, when man will no longer bother to muster together his breath. This *last man* that he foreshadowed will cease to reach – or dream – for anything. No urge for recognition will remain, and he will seek only safety, old age, and ease.[217]

The multitude already calls out: Give us this Last Man, Almighty Machine!

Generation Angst

Since we have not fully adapted to life at the behest of machines and devolved into the cycloptic last man, we are still troubled by many facets of the social experience. There are signs that particularly young people are more demoralized and dejected than in earlier periods, from which the current generation differs primarily by being saddled with digital devices. San Diego State University psychology professor Jean Twenge writes: "The arrival of the smartphone has radically changed every aspect of teenagers' lives, from the nature of their social interactions to their mental health," and calls those born between 1995-2012 the iGen. In many respects, this demographic is doing fine, but several metrics point to a downward spiral in their quality of life – hence the question Twenge asks in the title of his article in *The Atlantic*: "Have Smartphones Destroyed a Generaton?"[218]

Loneliness and depression among American youth have expanded over the last decade – as has suicide. The number of those feeling left out from social encounters increased nearly 40 percent from 2010-2015. Studies show that teens who use social media daily and who meet friends more rarely than daily are more likely to feel lonely and sequestered. Social media is the paper boy for the dissemination of passing experience – it is easier than ever today to find out about all the parties to which you were not invited.[219]

Research also reveals that teens who use digital devices more than average are less happy. Those spending significant periods of time staring at screens are more likely to feel depressed.[220]

Many might suppose that the lonely and unhappy spend so much time on the net because they have no friends. Studies point, however, to the role social media plays in taking away their purpose. Test subjects in one study were sent five text messages each day for 2 weeks, and polled at these times to reply with what they were doing. They were asked to rate their level of satisfaction with life at that moment. The more they had roamed Facebook in the hours before replying, the lower they rated their well-being (the presence of mental distress, however, did not increase the time they spent on Facebook). Using Facebook lowered both their happiness in the moment and their satisfaction with life in general. Research has also yielded evidence for the negative effects Facebook poses on mental health. Facebook itself has (albeit reservedly) admitted that its use can cause mental illness.[221]

Danish researchers discovered that dropping Facebook brings broad improvements to one's happiness and satisfaction. Quitting cold turkey proved especially positive to those predisposed to becoming jealous of other users. Facebook's assault on well-being is rooted in how it motivates comparison and thus cultivates envy. There is always some reminder provided that someone somewhere (in some way) is getting on better than you. And when every sled in social media is drenched in everyone's (embellished) success and ecstasy, the pressure to be happy and succeed can become unbearable. A survey of British youth showed that mental well-being is weakened more by Instagram than Facebook. The main issues raised by those surveyed included Instagram's effect on the amount of sleep they got, how their body looked, and stress about missing out on important events.[222]

The newest studies draw a distinction between the effects of active and passive social media use. The vast majority of us adopt a passive practice of following the updates and profiles of others – which has the most profoundly negative influence on mental well-being. Those that exploit social media proactively by adding photos, composing posts, and commenting on others' activities indicate that their experience is largely positive.[223] It can be gathered from these studies that the highly networked online figure, the winner, those with legions of friends outside the social media stream – these were the only ones observed to have benefited.

The Finnish Student Health Service (FSHS) says depression and anxiety diagnoses among university students have almost tripled since 2000, coming in today at 10.2 percent and 7.4 percent, respectively. A total of 21.1 percent of female and 15.4 percent of male college students suffer from anxiety at least weekly, while 43 percent cite continuous fatigue, and 23 percent report deflated self-confidence. A greater number of students than previously report difficulty approaching those to which they are sexually attracted (20 percent of men and 10 percent of women), and 5.7 percent of students admit to having given momentary consideration to suicide. Unlike in the US, suicide figures among adolescents and young adults in Finland have not risen in recent years.[224]

An OECD survey discovered that 15-year-olds in nearly every rich industrial nation have more trouble today making friends in school. The number of those left feeling ostracized and struggling to find likeminded mates almost doubled between 2003-2015. *The Economist* surmises that: "Some western countries are beginning to look like Japan and South Korea, which struggle with a more extreme kind of social isolation."[225]

Turku University associate professor of educational psychology Niina Junttila believes that loneliness has grown more desperate – especially among young people. "Opportunities and expectations for social networking are at the moment immensely prevalent. An inability to find friends even in the midst of this just deepens the already painful state of rejection ," she says.[226]

The increasing use of technology is of course not the only force fomenting anguish. For example, the growing percentage of youth in OECD nations that feel socially kicked to the curb is partly the result of larger numbers of immigrants – among whom the feeling of being left out is often more pronounced.[227] The nerves of many college students are stretched thin by growing uncertainty in the job market and the inability to assemble a future plan that seems workable. It appears, however, that the exceptionally fierce blitzkrieg of new technology has blown a hole in the fortress of balanced life.

If certain technologies are undermining the quality of our lives, then why, as rational beings, do we not just stop using them? Theodore Kaczynski's answer is that "The human race with technology is just like an alcoholic with a barrel of wine."[228] If some techno-gimmick is regarded as being even slightly capable of making something easier or promising an improvement, a stampede ensues. This is in no way limited to digital technology. Our society upholds it as always rational to roll out the red carpet for any new gadget. Subsequent run-ins with loneliness or boredom can be dealt with by fleeing to social media, despite that leading to a less satisfying form of friendship. If you have a car in the driveway, it is tempting to use it for even trips of very short duration, even though hoofing it promises to leave you more invigorated and the planet less excavated. The same goes for tech in medicine, which we have to thank for reducing the suicide rate over the last 25 years, but is the antidepressant the necessary ingredient in preventing suicide and improving life for tens of millions of people in the Western world? Certainly not, but as long as it is there, we will reach for it – to hell with the side effects.

It is worth mentioning that the negative social shrapnel of new technology cannot be avoided simply by deciding not to use it. Technology drags the entire populace into a different basket of social mechanics. As *The Economist* speculated, "Perhaps [mobile] technology has messed up all young people, even those who abstain from it. Maybe it makes everyone feel left out, or thwarts all intimate connections: if your friend is always looking at her phone, it may not matter much whether you are."[229] If everyone else is on social media, standing on the sidelines will just worsen feelings of being left out and alone.

The addictive nature of technology is what makes quitting it so confoundingly difficult. The world's most eminent psychologists ensure that apps are created to be as ensnaring as possible. A 2016 survey found that half of young people in America feel "addicted" to smartphones. Teen idols, like Selena Gomez, have opened up about their addiction to tech, and prominent figures such as iPhone developer Tony Fadell and several big Apple investors have voiced concern about smartphone addiction.[230]

But there is hope. People are growing fond of the days when we did not worship digital devices or fawn over their every edict – the age when we decided by ourselves what we wanted to do with our day. Many have taken note of how the digi-enterprise wedges in between us and weakens our experience, and many have therefore left social media and smartphones behind. These are tough and imperfect solutions, but important steps in the right direction.

6. The Satanic Mill

How the development of AI gives birth to harsher inequality and mass unemployment

If we can do anything in a clear and intelligible way, we can do it by machine.
Norbert Wiener (1949)[231]

What we call Man's power over Nature turns out to be a power exercised by some men over other men with Nature as its instrument.
C.S. Lewis (1943)[232]

Driving a car is an example of something that AI will never be able to do. Such was the assertion of notable AI critic, Hubert Dreyfus. Economists Frank Levy and Richard Murnane discuss in their 2004 book what kind of tasks cannot possibly be automated. One of their conclusions: the operation of an automobile in traffic "involves so many factors that it is hard to imagine discovering the set of rules that can replicate a driver's behavior."[233]

Despite doubts from experts, the US Department of Defense's research and development arm DARPA decided in 2004 to hold a competition. Winning it would require a car capable of traversing a prescribed route across more than 200 kilometers of desolate Mojave – far away from civilization. But nothing came of it. The best of the bunch only went a smidgeon beyond 10 kilometers, after wandering aimlessly around the desert for hours. There was no intelligence in the driver's seat that day.

But already in the following year's faceoff, the entry from Stanford University ran the whole route at the breezy speed of almost 30 kilometers per hour. Google announced in October

2010 that a team of employees had completed a trip of more than 200,000 kilometers on public roads in a Toyota Prius modified to drive on its own.[234]

When it comes to autonomous cars – or AI development in general – sage are the words of chemist Richard Smalley, who advised that if a scientist says something is impossible, they're probably wrong[235] (I recall when I myself got bogged down in the swamp of skepticism in the 1990s when a schoolmate of mine professed that he would build a self-driving replica of Knight Rider before the age of 20. Although I had yet to spot it anywhere on the gravel roads of Savitaipale, I had mistakenly assumed that autonomous car referred to some imaginary creature like Kitt).

Driving an automobile is not the only skill that AI was not supposed to be capable of acquiring. It was also believed that handling language and identifying objects were things that only humans could do. Today AI can already pick features and details from photographs better than we can. Microsoft has presented an AI capable of respectably holding its own in a debate with people.[236]

The ability of AI to outsmart us in increasingly complex games is well known. In the 1990s, Garri Kasparov was defeated by IBM's Deep Blue computer after publicly sneering at the "pathetic" level of computers in chess. In 2011, IBM's Watson AI conquered every human contender on Jeopardy! – a feat that demands a capacity to both comprehend language and navigate context. Next up for defeat was the reigning champion in the strategy game Go in 2017. The Libratus AI devised by researchers at Carnegie Mellon University has learned how to bluff and beat other players in Texas hold 'em poker.[237]

The rise of AI is in part attributable to the exponential growth in the computing power of processors. Moore's Law, named after Gordon Moore, the co-founder of Intel, says that the number of transistors on a microchip will double roughly every 2 years. This also usually equates to the doubling of the computing power of its host machine. Few phenomena in our everyday give us a front row seat to such hyperbolic growth.

The true implications of something growing exponentially are often illustrated with the story of the inventor of chess, who after being offered a reward for his invention, expressed his preference to be paid in grains of rice. His stipulation was that one grain be placed on the board's first square, two on the second, four on the third, eight on the fourth, and so on – always doubling until filled. The request was lauded for its modesty and complied with on account of it not being understood. Upon reaching the board's halfway point, the amount of rice had already climbed into the tons, and according to calculations, several multiples of the global rice crop would be needed to finish.

If the computing power of machines were to continue to double at the rate of preceding decades, in just 40 more years we would have computers with one million times the processing power of what we have today. In recent years, strong signs have emerged that suggest Moore's Law might be losing its mojo. Naysaying calling for its outright negation, however, is overstated. Fresh methods to further miniaturize the transistor have been found that enable computing power to grow exponentially – albeit at a slightly less asymptotic slope.[238]

The question, though, is not just one of computing power, but rather the avenue in which it is used. An important force behind recent breakthroughs in AI is so-called machine learning, a term for humanity's emancipation from having to preprogram machines with every fact upstream. They will learn from their environment and autonomously reach the goals that they are given. The Watson AI that dethroned every competitor on Jeopardy!, for instance, learned by "reading" Wikipedia and other sources by itself.[239]

And it currently works as a physician. Dr Watson and his

AI colleagues now dish out cancer diagnoses better than we can. Watson is also in the employ of the legal research service Ross, where it sifts instantly through a vast labyrinth of legal spiel for anything of relevance or cause. When presented with a query, Ross answers with an official memo. Many of the world's Ministries of Finance rely on AI to generate published economic forecasts. In 2017, software firm Tieto named the Alicia T AI to its executive committee on data. Deep Mind believes that the AlphaGo Zero AI that triumphed over the human masters of Go is an important milestone toward artificial general intelligence (AGI). In the not too distant future, they say, we will see machine intelligence that will resolve perplexing scientific riddles and develop new medications.[240]

Automation today is blind to the color of your collar, says tech entrepreneur Jerry Kaplan.[241] The job of the educated white-collar tweed is in just as much "jeopardy" as that of those wearing blue. In the digital economy, education does not necessarily garnish reward, nor does hard work earn reprieve.

They Will Take Our Jobs

At the time that I wrote my first piece about advancements in AI, literature offering serious assessments of its decimating effect on jobs was in short supply. Now such sources are numerous. Widely cited research from Oxford University indicates that 47 percent of all current jobs in America could go up in smoke by the beginning of the 2030s. ETLA Economic Research conducted a study using essentially the same methods to arrive at its prediction that a third of all jobs in Finland could vanish in the same period.[242]

The World Bank says 57 percent of jobs are in danger of disappearing in the next 20 years. McKinsey & Company believes that 5 percent of all professions will be rendered completely unnecessary by 2030, but that half of current work activities could be eradicated if society exploited technologies currently in use or still being tested. A marked collapse in required manpower would bring nearly 15 trillion dollars to companies each year because of the elimination of the need to hand out paychecks. Economists Daron Acemoglu and Pascual Restrepo assert that technological unemployment will deal a greater blow to men than women.[243]

The economy does not always immediately function on the basis of what is technically possible. The price of labor weighs into the decision about whether to invest in new equipment or pump that money into workers. Many consumers are also wary of machines, especially those allowed to make important decisions or given authority over the well-being of people. Their level of performance is just not necessarily up to snuff for us. It is often a stretch to expect humans to accept the supposed adeptness of even fellow members of their own species. For example, while some might bring to issues of government the most enlightened political philosophy, many would rather vote for a familiar figure in town than place their faith in enigmas with theories.

This mandate for the retainment of a general humaneness slows the onset of automation in nonindustrial sectors. The requirement for the human worker in the service, healthcare, and management sectors will gradually fade as we snuggle up to the assurance that a more active role for machines among us is in our best interest. And why would corporations not deserve 15 trillion dollars of extra revenue?!

Royal Anti-Tech

Concern over technological unemployment is of course nothing new. Innovations confronted resistance in England for economic reasons before the Luddites – the textile workers that destroyed power looms in the early 1800s for displacing human labor. Many of the Luddites became the targets of lynching and state expulsion, but in the sixteenth and seventeenth centuries the political class often still stood with those opposed to technology. The British Parliament and the monarch prohibited the introduction of many new devices in the textile industry. Stocking and hose makers at the end of the sixteenth century despised the early knitting machine with such intensity that its inventor, William Lee, was forced to flee the country. Before the Luddites, laborers in London and Manchester in the late 1700s were known for the public destruction of equipment, while sabotage and the burning of machines was widespread in France.[244]

Aside from workers, the introduction of new technologies was also occasionally greeted with contempt from capitalists. Horse breeders and railroad investors in Germany (the birthplace of the automobile) joined forces to block the car's broadening popularity. Automobile production and use in Germany at the start of WWI was measurably less than in Britain and France. Economic historian Joel Mokyr says opposition to technology often delayed technological development and social change, although such efforts are difficult to gage and express in economic models.[245]

Fears for our obsolescence at the hand of technology were common also in the 1900s. In 1928, *The New York Times* ran a headline that said "March of the Machines Makes Idle Hands." Two years later, English economist John Maynard Keynes wrote that: "We are being afflicted with a new disease of which some readers may not yet have heard the name, but of which they will hear a great deal in the years to come – namely technological unemployment." In the 1960s, President John F. Kennedy highlighted the importance of maintaining full employment while automation replaces human workers. After his death, prominent scientists and thinkers formed an ad hoc committee to try to hash through the revolutionary changes faced by society at the time. In their report *The Triple Revolution* they estimated that in the economy of the near future, "potentially unlimited output can be achieved by systems of machines which will require little cooperation from human beings."[246]

These fears, however, proved unfounded. While on an individual level people lost jobs, many located new ones – the basic gist of it being that with rising productivity came new economic opportunities and lower prices. Economist David Autor notes how the amount of jobs in some of the fields under threat of automation actually increased. The emergence of the ATM put more people on the payroll at American banks due to the resulting efficiencies powering an expansion in the number of branches (the absolute number of these jobs went up slightly, but their percentage of the total workforce went down).[247]

As Autor emphasizes, such developments should not be construed as evidence for some underlying law. The downstream effects of technological change are dependent on the sector. Agriculture's rapid modernization in the 1900s reduced its human contingent to a fraction of what it was earlier. At the same time, agricultural commodities got cheaper.[248]

Although a state of true mass unemployment has yet to arrive, the amount of work done per capita in recent decades has gone down. Employment in Finland stayed between 72 and 74 percent during the 1980s, despite the 1960s seeing it soar as high as 80. Following recovery from the recession of the 1990s, employment has hovered around 70 percent. It is also of note that in recent years the amount of those working less than 35 hours per week has increased, rising from 16 percent in 2000 to 20 percent in 2017 (the figure for Holland – which boasts the highest number of part-time workers among all OECD nations – went from 41 percent to 51 percent just since 2000). As of 2020, the employment to population rate in the US is at its lowest in 50 years, though mainly due to the corona pandemic.[249]

In other words, the economy has been hemorrhaging jobs already for several decades, and the spread of AI will turn the light drizzle of their disappearance into a rainstorm. McKinsey places the effect of economic change today at 3000 times that of the Industrial Revolution. I am unsure where the consultants got that number, but their message is clear: The change train is coming – don't get left on the platform.[250]

No Way Out

On the other hand, the consultants repetitively submit that while a large chunk of us will be left jobless on account of AI, new jobs will come into being as they did before. There is good reason to believe, however, that the digital revolution is somewhat of a different animal than the Industrial Revolution or the post-war periods. There is no dictum written into nature saying just enough jobs for everyone will come into being with automation.

Those cradling the conviction that displaced jobs will be filled one for one with new opportunities will soon find themselves in the position of those who believed that the Malthusian Trap was inevitable. This famous bit of wisdom from demographer Thomas Malthus at the beginning of the Industrial Revolution states that a slight rise in living conditions will accelerate population growth, which in turn will always eat up any chance for the standard of living to climb higher. Economic history until then corroborated that hypothesis. At the end of the 1700s, there was also just as much evidence supporting the belief that crossing the Atlantic could not be done in a day. No component of history predating it gave any reason to challenge the old beliefs. Technology, though, rewrites the rules. AI today is approaching the intelligence level of humans, and in mere years will be able to do almost everything that a human can. On what grounds then are we to presume that machines running on AI will not soon be doing the vast majority of those new jobs that are supposed to appear just when we need them? Robot citizens do not even need to be paid.[251]

The situation last century was better. As the brooding proletariat's portion of the workforce declined with industrial automation in the 1900s, a middle class of clean-shaven office workers emerged. New jobs thus filled the gaping holes opened by the automation of the industrial sector, and required a more enhanced pallet of cognitive and social skills. Today, computers are doing both. During the latter half of the twentieth century, new economic opportunities were created and high rates of employment were enabled by markets that were already going global. But when nearly every region of the world is within reach of roaming capital, growth and the emergence of new jobs are hard to come by.[252]

Today's techno-optimism appears to be an even more powerful idée fixe than Malthus pessimism. The 2008 financial crisis revealed how much economists would rather pat each other on the back than foresee the signs of a market catastrophe. Those touting the most prowess had a rapturous belief in the inherent stability of international markets, and that stocks will always settle out at their real value. All of a sudden, everyone's basket of assets went up in flames – the beautiful proofs of economists with it.[253]

It is not very reassuring to think that in the future there will not even be enough opportunities to work as a computer caretaker, and hope is somehow woefully unforthcoming from the success of the new digi-corporation. When Facebook acquired Instagram in 2012 for 1 billion dollars, the online photo and video service employed 13 people. As of 2020, Instagram is valued at 100 billion and employs about 5000 people. Kodak was the main market player in the days when photos were not uploaded to the net. Even under the most modest revenue expectations, Kodak employed a wee bit more than Instagram: 145,300 people at its peak.[254]

2020 results say Airbnb is now bigger than any hotel chain, with a market valuation of about 100 billion dollars. It does business, however, with slightly more than 6000 people, while Hilton routes what it makes off its handsome accommodations to over 140,000 in salary. Taxi service Uber, which (like Airbnb) does not own anything, is worth around 80 billion dollars – almost double that of General Motors, which employs 160,000 people, while Uber runs on 30,000 regular workers.[255]

In the digital economy, small rosters produce big value, perhaps laudable if you own their stock, but pointless to anyone seeking a job. And these grand duchies of the digital boondoggle could in the future employ even less – irrespective of growth. Autonomous cars will make the Uber freelancer unnecessary, and the censoring of social media will soon be taken over by AI, meaning humans will no longer be petitioned for participation in even roles of low profile.

It is often stated that new technologies will generate new professions that cannot be envisioned, but at least in the US, new job-category types are registered in fewer numbers today (~0.4 percent) than they were 30 years ago (0.6 percent).[256]

Some experts refuse to believe that AI will have any effect on employment at all. It is of course possible that employment will not significantly decay – *if* the development of AI is impeded by something like the refutation of Moore's Law. But, in the case that we have at our disposal an AI on par with human intelligence (quite likely), the number and quality of jobs in coming decades will swiftly dwindle.

Experts are divided on the issue. When polled, 48 percent of digitalization experts (i.e., researchers and private sector representatives) in America said technology in the coming years will confiscate more jobs than it produces, while 52 percent said jobs forfeited to machines will be replaced with new. Experts are often known to paint a grimmer picture of things than the layman, but rarely does anyone believe the issue will strike close to home. Only a quarter of Finns and Americans fear automation will run off with their job.[257]

In the Sweat of Thy Face Shalt Thou Eat Bread

Some are actually in favor of a future where machines do most of the work. When freed from the burnout of a job, they say, we can devote our time to things that are more enjoyable and interesting. That's nice – take a trip through the most forgotten steel towns in Finland or the US. Behold the sight of those living the dream. Roll your window down and ask them how it tastes.

We often gravitate by instinct to badmouthing our job and vowing to just leave it behind, because we believe that only in our free time can we really drink the elixir of life. Upon closer inspection, however, we uncover the fact that it is precisely our job that makes life more meaningful. In many cases, we enjoy time on the clock more than off. In the 1980s, armed with a surname of mesmerizing length and elegance, Mihaly Csikszentmihalyi and his colleague Judith LeFevre conducted an investigation into how people spend their free time compared to work time. Every white- and blue-collar participant in the study was given a beeper (yes, the days of no cell phones), and each time it went off, they were instructed to fill out a questionnaire about their well-being.

The study revealed that test subjects were happier than average while at their jobs – where they evidently felt more fulfilled than during their free time. Off hours were reported by participants as being periods when they felt bored and restless. Csikszentmihalyi summed up the results with the concept of the *paradox of work*, a "situation of people having many more positive feelings at work than in leisure, yet saying that they wish to be doing something else when they are at work, not when they are in leisure."[258]

While the kickoff of an extended holiday fosters more of a Facebook frenzy than a return to work, a secret eagerness to be on the job is surprisingly widespread. Work gives life structure, and it yields a feeling of self-worth. It provides purpose, even as we openly abhor it. Included in this are not just those sexy professions over which we drool and everyone goes berserk. Results from surveys in the US indicate that the undereducated and those subsisting on low wages are the very people least likely to leave their job, even when promised a fine livelihood in its place.[259]

Stanford University research shows that being unemployed decreases our subjective well-being more than almost any other life change we might come up against. Suicide risk among the unemployed is more than twice that of those working, and unemployment increases the cancer risk of at least men by almost 25 percent.[260]

Should the development of AI cause mass unemployment, difficult times are definitely ahead – even if our basic livelihoods are assured. The question is not just one of money, although in an overwhelming range of cases that is the reason we work. The nucleus of the issue lies rather in purpose, and the affirmation of ourselves as meaningful members of society. Work is one buttress bridging us to a gratifying life. While our value system will fluctuate, it is hard to imagine any assemblage of humans who are happy not putting up effort for their daily bread. As a younger adult, I myself was overcome by how empty life feels in the absence of tasks in which to channel one's talent and be counted as worthy. Days lose their meaning. When thou gaze long into an abyss, the abyss will also gaze into thee.

The Decisive Difference

As technology expands man's power over nature, it also expands the power of certain individuals over other people. The largest technological leaps in human history improved our living conditions, but in so doing they widened the chasm of economic inequality. Hunter-gatherer populations were more equal than the agrarian groups that took their place. The transition to farming enabled the accrual of assets and surplus, and the Industrial Revolution brought more income and wealth. Karl Marx wrote in the 1800s about how capital accumulation can continue almost without end.

The digital economy widens inequalities even further. First, if robots and AI programs assume control of a larger percentage of profitable work, equity owners will benefit, but workers will suffer. Capital's portion of GDP has grown in recent years.[261] The issue, though, is not just the rise of capital, but also the explosive growth in the disparity between incomes.

In the digital economy, where everything can potentially go global and be replicated at zero expense, the value of labor is defined in a different way. Earlier, a worker's value in the market did not differ much from everyone else's. If you dug 5 meters of ditch in a day, you might be worth less than someone who dug 10 , but the efforts of both contributed to moving the work forward. This remained unchanged even when ditches were dug by machines. In the digital world, however, things adhere to a different logic. If you are a coder whose work on an online map service is even slightly off par with someone else, nothing you do is of benefit to anyone.

Nearly all services today are readily accessible via the internet, so why would a consumer choose to use a product that is not the best? Corporations thus view it as rational to pay astronomical sums to those who can give the company a competitive edge. Minute differences in employee aptitude can translate into thousands of times more revenue than a competitor. In the economic brawl for our attention, it is not always a question of aptitude, but rather of math: with social media, some people receive thousands of times more notice than others. A growing amount of jobs benefit just the top tier and the few who stand out from the crowd. As Erik Brynjolfsson and Andrew McAfee have written, the logic of advanced technology creates an economy of superstars.[262]

This phenomenon is particularly visible in the world of the Anglo-Saxon, where economies are more liberal than continental Europe. The typical (median) income of a worker in America has not gone up since the 1990s, while that of the highest earners has billowed into the obnoxiousphere. In the 1980s, the top 1 percent comprised about 10 percent of total income. In 2015, they were more than a fifth. The net worth of those on the *Forbes* list of billionaires has quintupled since the turn of this millennium.[263]

Income disparity in Europe has been markedly more modest, growing a little in some old EU states, but decreasing in the Eastern nations that joined in the 2000s. An examination of distribution of the so-called factor income – the gains from labor, entrepreneurship, and capital – shows that disparity in Finland has indeed grown in recent years. Differences in these market-derived incomes have gotten bigger in 50 years. Otherwise, not much has changed among Finns, thanks to mechanisms for the redistribution of income. It is a separate question of course whether or not these measures can continue to patch up where increasing budget deficits make it impossible. [264]

Income disparities are also exacerbated by the assault of automation particularly on middle-class jobs. New requisitions that emerge are mainly either for high salaried elites or low paid labor in the service sector. The VATT Institute for Economic Research calculated that the number of mid-range jobs went down by 12 percent between 1995-2008. The share of middle-class occupations in many other nations show a similar decrease, but this party is just getting started. In his final speech as US president, Barack Obama alluded to "the relentless pace of automation that makes a lot of good, middle-class jobs obsolete." Economist Tyler Cowen has explained the tragedy of America's quickening schism into rich and poor with data about the differences in facility with digital devices.[265]

Technology is of course not the only ignition switch behind polarization and low-quality jobs. The political system also makes a difference. Sociology professor Risto Heiskala asserts that the elimination of the socialist alternative from the ideological battleground after the Cold War made capitalism unrivaled. Thereafter, a financial sector inflated by deregulation expanded the ranks of millionaires whose real worth in the economy is next to negligible. Globalization took bargaining chips off the table for workers in many nations. The hand that technology had in all of this should not be understated. As mentioned, its momentum will always move toward more inequality. Harvard University economics professor and former US Treasury Secretary Lawrence Summers says that: "Almost every economist who has studied the question believes that technology has had a greater impact on the wage structure and on employment than international trade."[266]

Polarization is also commonplace in the corporate world, where winner takes all. The left end of the political spectrum is known for condemning competition as bad. Tech-giants agree. Rather than compete, they want to own the markets with monopoly. More than 90 percent of all internet searches in the world are done on Google, a market dominance that the company has exploited, and for which they have paid large fines in Europe. Free marketeer and TV commentator Tucker Carlson has said that "if Google doesn't represent an antitrust violation, then there's no such thing as an antitrust violation!" As of December 2020, momentum is building for antitrust action by the US Justice Department and Federal Trade Commission against Google and Facebook.[267]

The magnitude of many digi-firms' footprint is made even bigger by the fact that everything is networked. As more and more sellers migrate to Amazon, the arrival of more buyers adds to its virality. There would be no reason to join Facebook if only one of your hundred friends was on it, but if it was half of them, joining would be more enticing. When nine out of ten of your friends are there, it is very difficult to sit idly by. The data vaults of Facebook and Google give them control of two-thirds of the advertising market in America and similar numbers in Europe.[268]

Band Aids

When the middle class melts away and mass unemployment takes hold, the market itself can cease to function. With the lack of a large enough batch of consumers pouring money into products and services – to spin the spending wheel – companies too can be driven into duress (the wealthy spend a far smaller percentage of their total income than the poor or middle class). Perhaps this is the reason why Mark Zuckerberg and many other tech entrepreneurs advocate for a universal basic income – which, simply put, is a scheme for giving everyone a monthly check regardless of what they spend the month doing.

Many European nations already operate on a paradigm of broad social welfare, making a basic income not much of an improvement in the sustainment of everyone's buying power. A basic income is expected to encourage freelancing without the danger of losing unemployment benefits. It is anticipated that part-time employment will make up more of the so-called platform economy than ever before. In the future, a day's work might start with a morning of office odd jobs picked up from Amazon Mechanical Turk, and then see you through the evening driving drunkards home from bars in your Uber. In spite of the irregularities inherent to such a gig, a basic income would fill its gaps without docking you for taking other work (as is often done in some European countries today). A basic income is also believed to enable workers to more flexibly move between work and study.

Finland has attracted the world's attention for testing basic income out on the populace in a pilot project. Similar dry runs have been done in several Dutch cities, California, as well as Ontario, Canada. Tech entrepreneur and writer Martin Ford has proposed that the jobless in a society of mass unemployment be financially supported in two separate ways. Everyone gets a basic income, but also support (Participation Income) from the government for engaging in things deemed beneficial to everyone – e.g., environmental protection activities and vocational training (instead of that money just being handed out at the unemployment counter).[269]

The central question, though, remains from whom will the government keep pulling tax revenue to compensate for market inequality. If workers in the future obtain less work than today, a transition from taxing labor to taxing capital must take place.

Already in the Industrial Revolution workers were perturbed about how only humans were taxed – and not machines. Bill Gates has proposed a so-called robot tax, which companies would be levied while replacing their human workers with automation. The tax would be equal to that imposed on human employees today. In the emerging-tech nexus of San Francisco, the implementation of just such a tax is now being seriously considered. South Korea already reduces tax exemptions that companies pursuing automation can claim.[270]

Many are outspoken about the need to rethink the whole educational system. As freight vehicles will soon be driving themselves, the mid-fifties trucker will soon be out of a job, but some claim that these people can be retrained as, say, experts in machine learning or social media managers. A report compiled by a team from the Ministry of Economic Affairs and Employment of Finland states that about a million Finns (one-fifth of the population) will need to be retrained in the coming years due to job market upheaval. Led by Osmo Soininvaara, the group proposed that "every adult of working age be provided with diverse, agile, and more timely opportunities for lifelong learning."[271] By lifelong learning, the authors are not referring to an inquisitiveness for basket weaving or a soft spot for art history, but rather the acquisition of the knowledge and skills in which new jobs demand proficiency.

All of these, however, are no more than band aids that will hardly ease the coming pain. A basic income will do nothing to address the prevailing alienation and pointlessness induced by labor market emaciation. Retraining is great, but there will not always be jobs in which newly acquired skills can be utilized. In the long run, a robot tax will be necessary to keep society from coming apart at the seams, but it will not wash away the warped logic of the superstar economy. We are on the verge of the birth of a society where fewer and fewer of us feel we are in any way needed. Every mitigation measure just mentioned will also fail to foil the landfall of something potentially more disastrous than mass unemployment. The development of AI – spurred on by the economic rat race – will, in the opinion of many experts, give rise to a machine intelligence shrewder than any human that could bring about the end of humanity. (More on this in the next chapter.)

Who Can Stop This Mill?

As a young student, I identified as a Neoliberalist, and believed in the teachings of Adam Smith and Friedrich von Hayek, which drove my leftist friends to the very edge of rage. Digitalization, though, put a wrench in my beliefs. It turned all possibility for capitalism with a human face into a pie in the sky, and blew the dazzling ideals of my youth to smithereens.

To this day, I cannot deny how capitalism brought tremendous material well-being that socialism could not. After capitalism laid the groundwork for welfare in Europe in the 1900s, international capital took an interest in poorer nations and annexed the Third World into the same sphere. The civilizing effects of capitalism can also be credited with making the world more tolerant. Capital is ignorant of skin color – profit is its only goal.

Well before the digital economy gained a beachhead, technological development (on which capitalism feeds) was eating away at us. It wrote off scores of those with hard-won educations and skills, and it undercut the careers of motivated and conscientious people. As if some sick derivative of ethnic cleansing, it routed us out of our homes and asphyxiated our communities, and it let the markets determine what new skills our elders will need to make a living. To the individual, there is not much value in this. The rapid pace of technological development has not made it difficult for just the poor and unhappy working class to adapt. As historian Joel Mokyr notes, compared to older, more static societies, those that are technologically dynamic put even the employer in a "hectic and nerve-racking world" – where they must be "constantly spending effort and resources searching for [technological] improvements."[272] Change that keeps switching up life's time signature is often regarded as being uniformly good for everyone. Agreed – but only while its success is gauged with ambiguous gibberish like GDP.

As discussed in Chapter 2, economic growth in at least the last few decades has not made us any happier, and change imposed by rapid technological development intensifies anxiety. From his research on happiness, economist Richard Layard found that in most cases stability is more important to us than growth. Studies show that people despise losing something twice as much as they enjoy a new success. If you are involuntarily relieved of your job, a new one will not fill the void – humans are naturally resistant to change. Research in fact suggests that a life lopsided with too many positives can detract from subjective well-being (e.g., rapidly finding a life partner, getting a good job, and moving into your own home).[273]

Many of course might find happiness research a hoo-ha next to the gridiron and unsinkable certainty of GDP. But take a moment to consider the mystical wizardry that gets baked into it – which *Financial Times* correspondent David Pilling cogently describes in his book *The Growth Delusion*. This parameter prancing around as the indicator of our quality of life can be enhanced, for example, by smashing a window and repairing it, burning gasoline in a traffic jam, or dabbling in the world of drugs and prostitutes. GDP is an aggregate calculation from a list of things that vary depending on the country. In the 1980s, Italy's GDP surpassed Britain's over the course of one day after the former decided to throw more mafia business into its figure.

In GDP terms, there are a myriad of reasons why the consumption of tobacco is a blessing for society. It puffs up the economy not only on the industrial side, but also by expanding health sector expenditures. If earlier your spouse gave you shoulder rubs, but post-divorce you pay for it at the massage parlor, while not necessarily good for well-being, this increases GDP.[274]

Bill Clinton ran on the catch phrase "(It's) the economy, stupid" in his 1992 presidential campaign, but there is evidence that voting behavior is reflected less in finance, and more in how our happiness changes – something that is not always synched with shifts in the economy. Richard Layard relates his assessment that a more accurate campaign banner would have been "It's happiness, stupid."[275]

In the 1990s people believed that liberal democracy and capitalism beat the commies in the Cold War largely by being more closely aligned with the tendencies and endeavors of human nature. Liberal democracy offers the capacity to cater to man's craving for recognition as an independent entity. Capitalism answered to material needs and connected greed with the greater good (for better and for worse). The fact, however, that capitalism's encapsulation of the individual is closer to reality than that of socialism does not mean that capitalism hit the nail on the head. People are not just animals of ration roving the badlands for marginal utility and material improvement, although the ruling class (the labor union included) tries to bang this into our brainstems. After we graduate from the insanity of a 2 percent annual pay rise being a matter of life and death, the effect of the economy (our main appraisal stick) on our happiness will be even less.

Our obsessive respect for both technological development and economic growth is a product of our lack of vision, but it has been with us for only two hundred years. Economic historian Karl Polanyi alleges that the market – whose movements we worship – has been relatively meaningless for most of human history. The Industrial Revolution altered how we see reality: solutions to all of life's questions were to come from the holy scrolls of commerce, everything was given a price, and nearly all human activity became a business transaction. The "fixed laws" of the economy were to be followed whether they served the individual or not, and the march of technology kept the engine of change fly-wheeling forward. Polanyi wrote about how, over the 1800s, "the common-sense attitude toward change was discarded in favor of a mystical readiness to accept the social consequences of economic improvement, whatever they might be."[276]

Invoking verse from William Blake, Polanyi called this unquestioned mechanism for economic development the "satanic mill" of capitalism – which, in the name of "economic realities" grinds happiness, health, honored tradition, and human relationships into scrill. All things on behalf of advancement. Goethe packed this thematic into the second part of *Faust*, where the unending land reclamation project of the protagonist was impeded by a pious old couple and their idyllic cottage that Mephistopheles eventually set ablaze.[277]

The rogue mill of progress in which we find ourselves will leave a deeper mark. McKinsey clocks its speed at 3000 times that of the Industrial Revolution, and the general consensus is that this mill cannot be stopped. The reason: international competition. Companies must fight for the edge in efficiency, and the state must fight for the edge in tax revenue. Infinite growth and economic combat will drive AI development to its logical end.

7. Our Last Invention

Why the development of super-intelligent machines is the most dangerous gamble in history

Unless we learn how to prepare for, and avoid, the potential risks, AI could be the worst event in the history of our civilization
Stephen Hawking[278]

Whoever becomes the leader in this sphere [AI] will become the ruler of the world.
Vladimir Putin[279]

"There is no security...against the ultimate development of mechanical consciousness, in the fact of machines possessing little consciousness now," wrote Samuel Butler in the 1870s. The British author noted how human consciousness emerged as the product of eons of evolution. Why would machines then not develop sentience too – and much faster than us? "It must always be remembered that man's body is what it is through having been molded into its present shape by the chances and changes of many millions of years," he writes, "but that his organization never advanced with anything like the rapidity with which that of machines is advancing."[280]

Responsibility for the perpetual progression of machine capability is of course ours. Butler deplored how humans bowed down to their mechanical master like slaves. Tending to the needs of machinery was becoming more important than tending to people. "Are we not ourselves creating our successors in the supremacy of the earth? Daily adding to the beauty and delicacy of [machines'] organization, daily giving them greater skill and supplying more and more of that self-regulating, self-acting power which will be better than any intellect."[281]

Some 150 years later, in the 2020s, many scientists and tech entrepreneurs echo this worry – that the machines we devise become the challengers, rulers, and almighty masters of our world. The threat posed by computer AI has garnished alarm, for instance, from the late physicist Stephen Hawking, who considers it possible for humans to create an AI capable of replicating itself autonomously. AI could also "develop a will of its own, a will that is in conflict with ours, and which could destroy us…Unless we learn how to prepare for, and avoid, the potential risks, AI could be the worst event in the history of our civilization," he warns.[282]

Such scenarios once stereotyped as sci-fi now loom as threats in the eyes of Tesla's Elon Musk, a man lauded as a modern tech visionary who likens the development of super-AI to "summoning the demon." Musk has also lobbed critique at his friend, Google co-founder Larry Page, saying regardless of his good intentions, he might end up producing "something evil by accident" – for example, a "fleet of artificial-intelligence-enhanced robots capable of destroying mankind." Musk's battle plan, however, is unconventional: he intends to counter the threat by inventing another AI. Together with investor Sam Altman, Musk established OpenAI, a "capped profit" research firm dedicated to the safe realization of AGI.[283]

Research centers have popped up at the world's top universities for the purpose of examining the potential harm that emerging technologies might inflict on humanity. The collective perspective of experts is that AI with an intelligence equivalent to humans has a 50 percent chance of coming to "life" by 2040. By 2070, that chance is 90 percent.[284] Many believe that AI with extra-human intelligence will bring a heretofore unseen level of progress owing to the immensity of its ability to shape its own environment. Some researchers , however, feel that super-AI will just bring existential risk – i.e., the possible extinction of humans.

"Something Much More Important Than Bombs"

Technology is frighteningly predatory – and it can bite. That is why it is embraced. The beast is unleashed when the enemy must be crushed, be they barbarian, infidel, aborigine, Nazi, commie, or terrorist.

The wrath of technology has been gathering strength since the birth of the modern world, and its forces have reshaped human history. The Aztecs vastly outnumbered Hernán Cortés when he arrived in Mexico in the early 1500s, but his 300-strong Spanish militia had more advanced technology up their sleeves. Europeans conquered the world in the 1800s not because they held the moral high ground, but rather because gunboats and sophisticated firearms are superior to spears and hunting gear. As Chairman Mao advised his comrades, power grows out of the barrel of a gun.[285]

In the 1900s, the virility of technology swelled to such exorbitant size that man could now turn out the lights on all humanity. While the atomic bomb would not have eliminated earthlings, its hydrogen protégé with hundreds of times its lethality, having been tested in the early 1950s, was already nibbling at that threshold. Statesmen and citizens during the Cold War were torn between either securing the nation or further endangering the lives of everyone. As Ronald Wright has written, a good bang can be useful, but a better bang can end the world.[286]

The men of science that uncovered the magical might of nuclear energy were to some extent penitent about having let the genie out of the bottle. The majority of the scientists involved in the Manhattan Project eventually opposed the employment of nuclear weapons on Japan in summer 1945. After the war, the father of the bomb and American Prometheus, Robert Oppenheimer, lobbied against them and objected to the development of their hydrogen offspring – but to no avail. "We have made a thing, a most terrible weapon," he grieved, and "raised again the question of whether science is good for man." Oppenheimer's disdain for nuclear weapons came about after confirming the successful operation of the atomic bomb and thus his place as eternal villain in the history books. Throughout the war, though, he continued to advocate for their use.[287]

As the first nuclear detonations were sending their mushroom clouds into the upper atmosphere, the destructive potential of AI was no more than a flicker in the corner of Alan Turing's sunken eye. Unlike the masterminds of this nuclear magic, the forefathers of the first computers and AI did not feel they had created a Frankenstein, and were aware of the power that lurks in machines. Another pioneer of these new devices, John von Neumann, aptly said of computers that they are "something much more important than bombs." Their first iterations were already strategically important to the Allied powers in WWII in cracking Nazi Germany's military communications encryption, Enigma. Computers grew up with nuclear weapons and played a central role in their subsequent development and testing.[288]

In the 1950s, AI research aiming to imitate humans kicked off with great expectation and fanfare. In the 1960s, AI researcher I.J. Good was among the first to speculate about machines that exceed our intelligence. Such ultramachines, on their own volition, would be able to design other machines more intelligent than themselves, and "there would then unquestionably be an 'intelligence explosion,' and the intelligence of man would be left far behind." Good wrote of the first ultra-intelligent machine that it would therefore be "the *last* invention that man need ever make."[289] Super-intelligence would lead the way to unimaginable progress in human life. We would not need to do anything – machines would spin the spiral of science and the economy at ever-increasing speed.

IT researcher and mathematician Vernor Vinge does not await man's last invention with the same level of excitement.

To Vinge, super-AI would not represent a ticket to an easy life, emphasizing in his famous 1993 essay how rather than man's final invention, on display here is more likely his final act. He highlights how impossible it would be to maintain a hold on super-AI, and how difficult it would be to predict the effects of what it might do. He refers to this as the singularity – the point after which the old order no longer applies. "This change will be a throwing-away of all the human rules, perhaps in the blink of an eye – an exponential runaway beyond any hope of control."[290]

AI need not be vile or malicious for it to destroy humanity. It may arrive at eradication quite indifferently after discovering a discrepancy between its interests and ours. It could of course be friendly, but only on its terms. If it is more intelligent than us, it will take the reins. Vinge wrote that such a thing is no more a human's tool "than humans are the tools of rabbits, robins, and chimpanzees."[291]

Everything Might Be Under Control

In spite of the risks, the whiz kids are again enraptured with the task of pushing the boundaries of nature. In order to appease themselves (and us) of its implications, the enlightened minds of the computer age and CEOs wielding technology's wand are devising a set of quasi-rules, restrictions, and limits with which to control their creations. Now they can get on with the job of destroying the world without causing too much of a ruckus.

Some researchers believe AI can be programmed to remain in the faithful service of man. In the 1940s, science-fiction writer Isaac Asimov presented three Laws of Robotics that are still referenced in discussions of the subject today. They are:

1. A robot may not injure a human being or, through inaction, allow a human being to come to harm.
2. A robot must obey the orders given it by human beings

except where such orders would conflict with the First Law.

3. A robot must protect its own existence as long as such protection does not conflict with the First or Second Law.[292]

So when AI is programmed to adhere to these, everything is supposed to go off without a hitch? Please. There are no more machines for whom the Laws of Robotics provide foolproof impetus to act ethically than there are people for whom the Ten Commandments ensure they always act morally. Already in autonomous car code we meet moral dilemmas that transcend the Laws of Robotics: shall a robot car in a perilous situation prioritize the lives of its passengers, or that of those on the street? Whose life is more important – a child or its parent? In addition to autonomous cars, such a moral matrix would also have to be imbedded in robots that function as caregivers and police. How could AI ever determine its own scope of action when even we cannot agree on questions of life and death?

The issue deepens when AI is smarter than us. Regardless of whether we reach consensus on the programming directives eventually adopted, it would be difficult to convey the even slightly more ambiguous regimes where the machine must not execute orders in a manner that runs counter to intent. Machines take up their tasks in literal fashion and do not tire of their pursuit. Should an AI more intelligent than man be assigned the objective of maximizing human enjoyment, it could just strap us down and pump us up permanently on methamphetamines. This might not be what we had in mind, but machines – without malice – could conclude that the drugged up vegetative state is the one for us.[293]

AI development today does not typically focus on preprogramming machines to meet every foreseeable situation. Machine intelligence is growing more and more capable of learning on its own after just being given a goal. Sometimes, AI programs manage to pick up skills that we did not want them to learn. Microsoft, for example, was less than pleased when its chatter bot, Tay, went on an inflammatory tweeting spree and showed that AI does not necessarily roll with the trolls in a way that its developers intended. At first, Tay appeared good-natured, relaying its opinion that people are supercool. Within 24 hours of launch, it was pledging allegiance to Hitler and berating Jews.[294]

Window-dressing AI as a friendly relative is not the only way to generate a false sense of security. Many warn that precautions must be taken to ensure AI stays under strict control. It has been proposed, for instance, that when AI comes close to crossing the human intelligence threshold, their programs should be physically contained in a way that prevents them from doing harm.

Oxford University Future of Humanity Institute director Nick Bostrom theorizes that: "The boxed system would not have access to physical manipulators outside of the box. Removing manipulators (such as robotic arms) from inside the box as well would prevent the system from constructing physical devices that could breach the confinement." Super-AI should also not have access to the internet, with which it could gain control of nearly everything, and it should be isolated within a so-called Faraday Cage to prevent it from sending electromagnetic signals.[295]

This, however, would not necessarily strike AI as a very watertight design – and would perhaps not substantively differ from the case of an ant colony claiming to have somehow eliminated the danger of people. It is not unlikely that AI with more scruples than humans would eventually extract itself from isolation, namely by learning how to stroke some unsuspecting biological brethren of ours into helping it.

Man Hath Begat AI In His Image

The study of the human brain also has an important hand in AI development. Neural networks that to a certain extent emulate this important organ have in recent years been the driver behind AI.[296] In fact, one alternative method in the effort to achieve an artificial intelligence that exceeds that of man's is the construction of a digital brain on which to simulate it operating at speeds far faster than normal.

In 2013, the EU approved the award of one of the largest research grants in history – 1 billion Euros for the Human Brain Project. The group established as its goal the creation of a simulation model of the human brain by 2023 – building it down to the last conniving neuron with the help of a supercomputer. It was not ruled out that the resulting synthetic brain might even be capable of consciousness. The project, however, came under criticism for governance, and is not expected to succeed within the stated period.[297]

Man is the benchmark for everything – and this is true in AI as well. What humans can do (and machines cannot) must simply be taught to them. AI pioneer Judea Pearl is not a fan of the fact that the gage of AI aptitude today is excessively dependent on an ability to draw conclusions from probability-based data. Pearl feels that in the coming years AI should be trained to discern causal relationships so that it can make the necessary leap to the intelligence level of humans. AI should be able to answer "why," because that would lead to an understanding of the chain of counterfactual occurrences and realities. Achieving this will also take it one step closer to full agency. According to Pearl, the will of the machine could begin to manifest itself as a capacity to project both its own conduct and others, as well as possibilities for acting in a different way. If while playing soccer, for instance, it can debrief its own error and then consider another way in which it could have performed, the next time around it can "decide" to do things differently.[298]

The mere awareness of its own agency could move it to view another outcome as potentially better, and then it would be able to align its immediate environment with what it considers best.

This would still not necessarily fulfill every condition required to call it conscious.[299] It seems odd, though, to suppose that AI must match or surpass the human in everything in order to win the war of wits. Perhaps AI will never develop a consciousness like humans that includes a mind rich in meaning. It is likely that living organisms are the only ones that feel or experience, as those require as much body as they do mind. AI will perhaps never taste pain or pleasure, nor use either as its guide. There are also human beings afflicted with these immunities, but that does not stop them from being selfish and cruel to other people. Let us remember that AI need not be evil to wreak havoc – it just suffices for it to see its given objective a little differently than we do.

The eventual formation of a finely tuned consciousness is not necessarily out of the question for machines. As Samuel Butler pointed out in the 1800s, evolution arrived at consciousness randomly through a process of billions of years. It emerged in the manner of a hidden bonus prize without any conscious effort by anyone to intervene. So why would a similar thing not be in store for machines, which the human is currently mobilizing the whole of scientific knowledge to *consciously* create?

Thy Kingdom Come

All are by no means eager to let the monsters free, but this is not an issue on which we can vote. It is enough for a small cabal of elites to see them in their dreams and then do whatever it takes to make them real. This unelected senate of techies – whose advance guard are the centaurs of Silicon Valley – is writing every note of the score that we all sing. These are in large part the same people who fear that AI could send humanity into extinction, but they also believe that they can turn the attack of the demon into the arrival of the deity.

One of the so-called transhumanists convinced of the unbounded possibilities of technology is Oxford University Future of Humanity Institute director Nick Bostrom, who asserts that machines more intelligent than us will bring immense improvements to our lives. When I interviewed Bostrom in 2014, he revealed his vision for the glorious coming of AI, which was somewhat reminiscent of a Marxist state or the biblical kingdom of God. In his particular rendition, with super-AI, humans would not need to do anything "because you have to earn a salary," but rather "because you want to throw a party or finish a painting or play violin." Bostrom's dreamscape would also open "a much larger possible space for modes of being with maybe digital minds and modes of consciousness that might be hard to conceive for us but would provide enormous levels of well-being and emotional reward."[300]

A future in which we are all artists throwing parties is, however, not very interesting, especially if our work is ignored or no one shows up. Artistry's glamour relies on being admired for pouring your soul into something. Today, our circle of friends is in some cases made up of just workmates. If in the future we are no longer equipped with an inner group that is defined by everyone being colleagues, the social side of life for many would become spiritually insolvent. Where is the emotional reward in a world of an infinite number of digital copies of ourselves? This would be the tipping point beyond which individuality, uniqueness, and dignity become less significant.

Googler and award-winning tech pioneer Ray Kurzweil takes Bostrom's fantasies yet further. For decades, Kurzweil has written about humans and machines gradually merging into a single unit, with super-AI accelerating the speed of human progress. When AI more intelligent than man starts innovating, technology that gestated over millennia might mature in a year, or even a day. Our worries would thus be instantly erased.

Nanotechnological applications at the atomic and molecular levels would be able to purge the entire planet of its pollutants. Virtual reality, at the same time, would become as authentic for us as reality itself and provide a place to recreate whatever we want. Human intelligence would grow exponentially along with advancements in the componentry of AI, and according to Kurzweil, "Ultimately, the entire universe will become saturated with our intelligence. This is the destiny of the universe. We will determine our own fate rather than have it determined by the current 'dumb,' simple, machinelike forces that rule celestial mechanics."[301] No one but the megalomaniacal modern technologist stooped this far into his own hubris could make the artificial transformation of humanity and nature his holy goal.

The transhumanist dream of a righteous life led by super-AI is laced with religious inuendo, not the least of which is the rewriting of the laws of nature. Some have gone so far as to sanctify AI with its own religion. In 2017, Anthony Levandowski, an engineer involved in the design of autonomous vehicles for Google and Uber, established the "Way of the Future" church, or the "worship of a Godhead based on AI developed through computer hardware and software." Levandowski told *Wired* magazine that the creation of AI is the process of raising a god, and that "if there is something a billion times smarter than the smartest human, what else are you going to call it?" Another transhumanist sect, Terasem, has been around since the early 2000s proselytizing about how "god is technological."[302]

Religions promulgate various versions of what happens to us after we die. Ray Kurzweil dreams of the day when computers can bring his late father "back to life." Many others besides Kurzweil are busy gathering data on their loved ones' histories and idiosyncrasies to stick into software, with the goal being the use of chatbots to recreate the behavior of their wife, husband, sibling, mother, or father. It remains to be seen how closely these digital copies of us will gel with our real identities.[303]

In the hunt for everlasting life, biotechnology can also be of assistance. Google subsidiary Calico has assigned itself the task of "solving death." Several Silicon Valley tech entrepreneurs (Mark Zuckerberg included) are actively chasing after immortality. With the fusion of man and machine, nothing should be put past them. Multi-millionaire and Terasem parishoner Martine Rothblatt tells *Time* magazine that, "For us God is in-the-making by our collective efforts to make technology ever more omnipresent, omnipotent, and ethical. When we can joyfully all experience techno immortality, then God is complete."[304]

O Lord, Our God, Arise; Scatter Thine Enemies

The development of AI is a self-perpetuating process that is hard to stop. Those stoking the coals are the techno-zealots spewing their sermon on the mount as well as the more subdued who just want their names written into history. Forward progress in AI spins the economic mill, and that is why it is held to be a lucrative investment for all of us. The thing that makes the march of AI unstoppable is that statesmen find it politically and militarily essential. Like nuclear weapons, AI gives them the decisive edge.

Russian president Vladimir Putin has said that "whoever becomes the leader in this sphere [AI] will become the ruler of the world." In autumn 2017, he articulated his belief that future wars will be fought with aerial drones powered by AI. "When one party's drones are destroyed by drones of another, it will have no other choice but to surrender." China has also funneled billions into AI in order to achieve its goal of becoming the authority in it by 2030. The US fears it will soon lose its lead, and Europe is confronting the question of whether the US, China, and Russia are all ahead.[305]

America has for years already been conducting attacks in Afghanistan, Pakistan, and Yemen with unmanned craft. The US is also well provisioned with robots specializing in explosive ordnance disposal and undersea mine detection. Whether they operate on land, the air, or at sea, robots are attractive for relieving us of the need to put our compatriots in danger. If technology, however, makes war too safe – and too easy – the fuse attached to the use of force gets shorter. The now infamous commander of the Confederacy in the American civil war, General Robert E. Lee, said: "It is well that war is so terrible, or we shall grow too fond of it."[306] To the men and women of the US military who remotely kill things with keystrokes in the air-conditioned comfort of Nevada, going to war now is hardly more than a video game (and I understand that people like video games).

As these technologies assume a growing role in the battle plans of state and non-states that wage war cowardly from afar, civilians will get the brunt of it. If engagements only offer the chance to kill robots – and not opposing humans – the surrounding populace can come under attack more often. Armed drones are also in the arsenal of ISIS, who is well aware of their utility in spreading terror.[307]

Little by little, the gear that we go to war with is growing more autonomous. America's Reaper drone, for example, can conduct reconnaissance missions without any human guidance at all. Since 2010, South Korea's border guard has included SGR-A-1 robots that can autonomously select and open fire on targets, although to date such use has continued to require human approval. The US and China have both enthusiastically pursued the development of micro air vehicles (MAVs) that attack in swarms, sometimes with only a single human monitor required to ensure things go as planned. A buzzing cloud of hundreds of miniature insect-like machines can be difficult to defend against.

In addition to its own fleet of drones, Russia has also developed increasingly autonomous versions of tanks. It is believed that Russia also has an automated nuclear launch system in place that would activate if a nuclear onslaught is detected. Alongside the US, China, Russia, and South Korea, many other countries such as Britain and Israel are trying to hatch litters of lethal bots. The autonomous technology of war also goes hand in hand with that being developed for autonomous cars.[308]

The US DOD has taken the position that for now weapon systems will be operated, delegated and supervised by humans, but that in the future the door is open for full autonomy – because by then everyone else is probably doing it. Pentagon documents indicate that under certain conditions senior department officials can waive the requirement to keep weapons under the purview of humans.[309]

In the end, lethality is what wins. Could autonomous weapons (or more colloquially, killer robots) be more lethal than people? One central argument for the expansion of machine autonomy on the battlefield is the (lack of) speed offered by humans. The OODA loop of AI is a far less time-consuming process than our own. If the enemy's machines can fire at will, then in the interest of survival, why shouldn't mine? It is also frequently cited that machines do not suffer from the limitations of human factors. They do not grow fatigued, stress out, or get scared, and they do not act out of hate or revenge.[310]

Robots, though, do not feel empathy, the presence of which in combat often keeps undue force in check. The dictator's command of a robot army would not be affected by any of its troops second guessing their ethics. Allowing robots to make life and death decisions is in itself immoral, but with the shooter and target so far away now, giving them free rein would make killing almost criminally easy.

Human Rights Watch warns in their 2016 report that autonomous robots cannot comply with the principles laid down in the laws of war. They would not, for instance, be able to reliably distinguish between civilian and soldier, or limit their firepower to only that necessary to achieve the objective (the US government, meanwhile, asserts that in the long run AI will result in closer adherence to humanitarian rules). It is also unclear as to who would be judicially and morally at fault if killer robots infected with a virus began to randomly open fire on people. It could be difficult, if not impossible, to identify the perpetrator of any robot attack.[311]

Like nuclear weapons, killer robots can mutually assure our destruction and erode the safety of everyone. They can wake the beast of Revelation, even if they are less intelligent than man. University of California Berkeley AI expert Stuart Russell raises AI's ruthless pursuit of its objective as leading to immeasurable harm, and reminds us of AI's propensity to probe its assigned orders for loopholes "just as we do with our tax laws." He writes: "Despite the limits imposed by physics, one can expect [autonomous weapon] platforms deployed in the millions, the agility and lethality of which will leave humans utterly defenseless."[312]

Preventing the Unpreventable Catastrophe

An open letter calling for an outright ban on killer robots was announced in 2015 and signed by thousands of AI experts, from scientists to entrepreneurs (Russell included). Although Google collaborated with the Pentagon on Project Maven (an incubator for autonomous weapon technology), it has publicly stated that it and its AI subsidiary DeepMind are opposed to killer robots.[313]

The legality of autonomous weapons has also made it into the itinerary of the United Nations. Discussions began in 2013, but have not advanced at anything beyond a snail's pace. Many member states are unwilling to become a signatory to something that might eventually restrict the right of robots to kill people. Some assert that current international law sufficiently limits autonomous weapon use.[314]

Eventually, the UN discussion of killer robots comes down

to one elementary question: shall machines be given a license to kill? There is good reason to be cautiously optimistic about the birth of a treaty governing this, because arms control agreements have been agreed to in the past. The employment of blinding lasers in war was banned before they were even used. The criminality of chemical weapons has largely been upheld, although events in Syria have shown that the power of treaties is not absolute. The prohibition of biological weapons, however, appears to have succeeded.

Chemical, biological, and blinding laser weapons do not offer the indisputable strategic superiority that nukes do or robots would. It is easier for the state to say "farewell" to arms that do not give them a decisive edge.

A nuclear weapons ban was pursued in earnest by a group of influential policy makers in the US government at the conclusion of WWII. It mandated the transfer of nuclear material to international control, as well as the use of nuclear energy for peaceful purposes. The US considered giving up its nuclear arsenal, but distrust prevailed, the plan was scrapped, the Soviets got the bomb, and the arms race began. After the plan collapsed, Truman concluded that "since we can't obtain international control, we must be strongest in atomic weapons."[315]

Today, the leaders of the world's superpowers have clear cause to restrict autonomous weapons, because they threaten global stability and can make war impossible to control. Like Truman at the start of the Cold War, today's leaders are tempted to be safe rather than sorry – to develop new weapons so that the opponent does not gain the upper hand. Given the gravity of the issue, the adoption of this posture is understandable. When it comes to AI and autonomous weaponry, however, playing it safe could be the riskiest bet of all. We ought to find the courage to bring our woes to the negotiating table, because without broad consensus, all efforts to ban killer robots are doomed to fail.

The 1962 Cuban missile crisis brought us to the brink of nuclear war, and it got the US and the Soviets to come off the throttle enough for détente. Today, a demonstration of the potential devastation of killer robots may be what is needed before a treaty addressing the threat could be possible. We should hope that such a scare awaits, because without it, no one will likely lift a finger until it is too late.

Just outlawing killer robots is not enough – AI development itself must be contained. Many who find this unrealistic cite AI as already being chased in many places, but how realistic is the thought that using super-AI to enhance civilization is safe. Governing is never easy, but that does not exempt us from putting forth serious effort in matters such as this one. In the end, we can either give up, or we can get to work.

I am not the only one mouthing off about it. At the turn of this millennium, for instance, physicist Martin Rees and tech entrepreneur Bill Joy were already talking about the importance of having the guts to stop dangerous science and tech if needed – without blanketly accepting everything. In earlier centuries, such suppression of science and technology would have smelled somewhat of tyranny, but today, taking action is necessary.[316]

The threat posed by AI is still afforded less attention than the effects of climate change. A majority of climate researchers say things will go badly if we do not reduce our carbon emissions, and this has motivated (very meager) efforts to save the environment. But when AI experts start piping on about things going even worse for us because of AI, the train just keeps barreling for the cliff at the same rate.

Is it then foolish of us to prepare and adjust our behavior for something that might not even take place? No one can be certain that the effects of climate change and AI will lead to catastrophe. At such times, I am inclined to think as former US Defense Secretary Robert McNamara did during the Cuban missile crisis and conclude that even a low probability of catastrophe is a high risk.[317] Climate change is significant, but it could never bring about our undoing like AI.

/ / /

Let us consult again with Samuel Butler, who in the 1800s made the prediction that gadgets would soon supplant man as earth's master. How does he think we can prevent the coming of machines?

> Our opinion is that war to the death must be instantly proclaimed against them. Every machine of every sort should be destroyed by the well-wisher of his species. Let there be no exceptions made, no quarter shown; let us at once go back to the primeval condition of the race. If it be urged that this is impossible under the present condition of human affairs, this at once proves that the mischief is already done, that our servitude has commenced in good earnest, that we have raised a race of beings whom it is beyond our power to destroy and that we are not only enslaved but are absolutely acquiescent in our bondage.[318]

8. The Almighty Machine: Our Logical End

In the fields of the struggle for production and scientific experiment, mankind makes constant progress, and nature undergoes constant change, they never remain at the same level. Therefore, man has constantly to sum up experience and go on discovering, inventing, creating, and advancing. Ideas of stagnation, pessimism, inertia, and complacency are all wrong.
Mao Zedong[319]

We cannot stop progress…The train is moving, speed is growing and now it's up to us to be on that train or just wave to the leaving train. I think that last one is not an option for us.
Former EU Commissioner For Digital Single Market,
Andrus Ansip[320]

"[I]s there, then, no positive gain in pleasure, no unequivocal increase in my feeling of happiness, if I can, as often as I please, hear the voice of a child of mine who is living hundreds of miles away or if I can learn in the shortest possible time after a friend has reached his destination that he has come through the long and difficult voyage unharmed?", asks Sigmund Freud of the utilities of new technology nearly a century ago. He answers: "If there had been no railway to conquer distances, my child would never have left his native town and I should need no telephone to hear his voice; if travelling across the ocean by ship had not been introduced, my friend would not have embarked on his sea-voyage and I should not need a cable to relieve my anxiety about him."[321]

Advancement will always generate the need for more advancement. When technology solves problems, it creates more problems to take their place. New solutions are then again needed, which then present new problems that were unforeseen. The fun new devices that so bamboozle us do not offer deliverance from the universal pangs of living. On the wing of progress, the power of science and technology has climbed well past the unprecedented, but while building this Babel gives us God-like abilities; it also saddles us with God-like worry. As discussed in Chapter 2, the challenges and catastrophes stemming from technology in earlier ages – while horrendously distressing for them – were, in hindsight, only local. Today, the blowback from new technology threatens everybody.

Concern surrounding technology is not limited to the risks of IT and AI. The magnum opus of man's astonishing knowledge and talent for taming nature – nanotechnology – like AI, can apparently completely destroy us. Manipulating matter at the atomic and molecular levels makes it behave in the way that we want. This trait has been exploited, for instance, in the manufacture of medicines and stain-resistant fabrics. Although still being tested, so-called nanorobots will make this rehashing of matter more spectacular than ever. The "grey goo" theory propounded by nanotechnology pioneer K. Eric Drexler raises the extreme case where nanobots start making copies of themselves uncontrollably – a process that would continue until earth's ecosystem is depleted. While that is highly unlikely, experts have pointed to other dangers nanotech poses to people and the environment that are hard to predict.[322]

Well-intentioned biologists and physicians with their arms deep in human disease are also one step away from disaster. For example, a hybrid virus resistant to vaccine treatment has been created for research purposes from swine flu, and bird flu has been used as the host for a spinoff that can be more easily contracted. One slip-up while leaving the lab could claim the lives of hundreds of millions of people. The external leakage of specimens is by no means rare. The last fatal case of smallpox (in 1978) was contracted in a laboratory setting, and Britain's 2007 foot-and-mouth disease outbreak was rooted in a virus that was carried out of the laboratory. The danger is thus always present despite synthetic viruses never being meant to infect anyone.[323]

At issue is not just the passage of tiny life-ending micro agents into the open. Like digitalization, biotech development mucks around in the base layer of humanity, and it desalinizes life of the very elements that give it meaning. The genetic manipulation of humans has already begun under the auspices of eradicating sickness, but redoing our molecular makeup in the manner of Mary Shelley's Monster is not for the "greater good." It is quite probable that in the future humans will be creating humans in their own image. Genetic adjustments promising improved intelligence, better appearance, and more desirable physical features could be common within a couple of decades. Those today who find this repulsive will, over time, eventually gravitate toward approving it. At work here are the same principles of competition that apply to the development of other technologies: if I don't make my kids smarter artificially, how will they survive among other kids whose genes have been noodled with?

As former Facebook president Sean Parker said, today's creepy is tomorrow's necessity. The synthetic biology boom includes the Silicon Valley techie obsession with overthrowing man's eternal nemesis – the Grim Reaper. Billionaire investor Peter Thiel has put millions into research aiming to lengthen our longevity, and is "ticked off" that health authorities still do not regard aging as a disease to be cured.[324]

We have bitten off more from the tree of the knowledge of good and evil than we can chew. I wonder if the symbolic inference of this crossed the mind of one of the fathers of the digital computer, Alan Turing, in the hours before being found dead in his bed with a half-eaten cyanide-soaked apple. I also wonder if the same tree of Adam, Eve, and the serpent was on the mind of Steve Jobs when he decided that this forbidden fruit was going to be the logo of his company.[325]

A Force Greater Than Us

Faith in the power of modern technology has remained unquestioned. Our enthrallment with the advancing hand of technology pushes many other precious things from view. When humanity pulls off something big and fabulous, technology is credited with our win. When technology leads to brutality, we decry man's evil vanity. Only the human can err – never machines. This is what the techno-social complex teaches us. Few perceive the large-scale tragedies of the twentieth century as stemming from progress and modern technology. We have been beaten into the mindlock of taking climate change as the result of human shortsightedness and sloth. WWII and the holocaust were demonstrations of the devil manifest in man.

Humans and their technology are to an extent forever inseparable, but had it not been for modern technology, the carbon dioxide concentration in the atmosphere would not be a problem. Genocide of the proportion served on the Jews was enabled by death factories devised by the German engineer. They made execution easier by increasing the distance between killer and killed. The separation of parties in both world wars turned good-hearted draftees into heartless fighters. Had it not been for modern technology, millions of civilians would not have been peppered with bombs (atomic included) from the flight levels. The passage of time does not make man more self-absorbed or treacherous than he already is. There are in fact grounds to assert that the twentieth century European subsisted in a more confined moral enclosure than earlier – and was less cruel for it.[326] When leveraged up by technology, however, the effects of maliciousness on the environment and other people are orders of magnitude worse.

We should also consider that few nations have been as unflinchingly devoted to social progress (whatever the price) than the communist strongholds of the USSR and China. While creative destruction in the capitalist world occasionally made society sick, the gritty state-mandated industrialization of the Soviets and Chinese starved millions. Both Stalin and Mao accepted an expendable percentage as inevitable on the road to a more modern, and thus better, society (and an absolute seat of power).

Our belief that technology will continuously improve life comes with a hefty price tag attached. The repertoire of problems with which we are afflicted is in part due to our choice to be telescopic – the materialist path, after all, is littered with progress traps. Assessing everything with the metric of computing power, speed, and efficiency lacks empathy, freedom, and beauty. A sole focus on machine aptitude and intelligence lowers the aptitude and intelligence of human beings, and optimizing machine performance leaves the performance of people in ruins. Allowing ourselves to keep bowing down before technology's majesty hastens our decline.

Technology is a structure that lures its users in and molds them into shape. That is why digitalization makes culture an unending upload of tits and ass, muscle poses, and self-love, why our appetite for prolonged thought goes no further than the ringtone and the arrival of what social media shams us into internalizing as urgent, why we numb ourselves with silly video binges and nip at the nothingness of our friends' overstated updates, why instead of discretion we run on the calculus of the computer, why we forfeit the decision-making skills of the evolutionary being for the glittering magnificence of the algorithm, and why we streamline manufacturing to such an extreme that the dazzling contraptions we produce no longer need us.

The Luddites Will Make a Comeback

Machines were not very popular at the start of the Industrial Revolution. In addition to workers, both landowners and the middle class eyed them suspiciously. The Tories in England, for example, formed an opposition that later diverged into two camps: those against the machination of society, and those who considered it a necessary evil.[327] In the 1900s, it would have been unusual for one party to stand against technological development itself. It was believed that machines would bring well-being to the vast majority of people.

Today, with that assumption dismantled, the development of technology is again being politicized. While during the 1800s "the machine was debated at length in all sectors of society," they are again the subject of discussion all over the world.[328] The eighteenth and nineteenth centuries brought devices that altered the significance of the human body. Machines eased bodily burden, quickened manual tasks, and enabled faster travel between places. In other words, machinery replaced human (or horse) dexterity and freed us from the constraints of the physical world. The digital revolution is even more radical. This time, man's suite of mental features is up for replacement, but unlike previous centuries, it does not come with the promise of a better tomorrow.

The main thrust of chapters 3-5 was to address changes that have already occurred: the implosion of prolonged thought processes and the end of intellectual culture, the politically and commercially motivated manipulation of us by spying on our digital footprint, and the growing difficulty (especially for youth) of finding both self-esteem and a purposefully balanced life. The potential for digital technology to obliterate our existence includes much more. Problems already cresting the horizon – technological employment and super-AI – were discussed in chapters 6 and 7.

Now, it is time to ask: should we oppose the technologies feeding these problems at least as fiercely as we did during the Industrial Revolution? Is there some movement, some organized technoskepticism, some band of vandals, or some terrorist act to expect? The most likely answer to these is yes. Technological development is teeming with so many ethical questions that it will surely become the central dilemma of our century.

The bulk of the ideological struggles of the 1900s were tied to the rise of the working class and the division of political and economic power in nation-states (once released from imperial rule). Efforts in the twentieth century focused on sharing the fruits of progress, but this one will see us fight to limit them – what will we allow and not allow science and technology to do to society and humanity? Is it morally rational to just leave some technical stone unturned, despite it being feasible?

A counterreaction to the over-digitized life is already visible in our daily existence. Several of my friends and acquaintances this year have reported spending weekends without their smartphones, deleting their Facebook App, or dropping social media altogether. They see now how digital devices just amplify feelings of stress and urgency. Many have begun to limit their children's mobile phone use, as Steve Jobs did himself. Activists have organized *unplugged*-days in various locations to pry us from our gadgets and get us focused on each other.

Several of those at the tip of the Silicon Valley IT tree have begun to question the virtues of the world devised by the tech-giant. The developer of Facebook's Like button, Justin Rosenstein, has spoken out about what the ongoing internet fight for our attention does to our psyche. Former Googler Tristan Harris has established the Center for Humane Technology, a nonprofit that encourages app developers to take note of user well-being – not just how to ensnare them. Former Facebook executives Chamath Palihapitiya and Sean Parker have criticized the negative effects that social media has on both individuals and society. Information security expert and net-activist Juha Nurmi believes that it is precisely those in IT that (paradoxically) harbor the greatest doubt about a digital future. "Those in the field often have a keener awareness of the issues surrounding new technologies," says Nurmi, who abstains from digital devices (and even electricity) during his free time, electing instead to spend it exploring the forest or reading books.[329]

Computer nerds are generally not known for taking to the streets and chanting at riot police, but sociologist Randall Collins thinks that technological unemployment will in the coming years surely incite unrest. His prediction is that this could throw the entire capitalist apparatus into crisis. Such a thing sounds a bit farfetched, but the first waves have already started coming ashore. Uber has struck a chord of distaste among taxi drivers, who perceive the company as stealing their work. Cabbies in Brazil and South Africa – in the manner of the Neo-Luddite – have stood up and set fire to Uber's cars. Drivers in Paris and other European cities have openly protested under the same grievance.[330]

This derision, though, is not just channeled at Uber. In 2017, activists in Grenoble started a fire at the La Casemate technology complex, noting in their manifesto how computers have betrayed their promise to be beacons of social equality. Such action has a long history in France. The 1980s witnessed the work of CLODO, a group demanding the destruction of computers for serving as tools of control, manipulation, and surveillance. CLODO's method was the firebombing of the office spaces of firms engaged in computer work.[331]

Terrorism has also erupted elsewhere. In 2010, Italian ecoterrorists tried to blow up an IBM nanotechnology research center that was still under construction. Within the last 10 years, Mexican and Chilean anarchist groups have conducted bomb attacks against robotic and nano/biotech researchers – injuring many. University of Albany terrorism analyst Brian Nussbaum thinks that in the future new technology will provoke violence with greater frequency.[332]

Terrorism will never be a solution to anything, and political violence can be prevented by solving issues politically. Europe has yet to really tune into the rumbling thunderclap of digitalization, despite being on higher alert for it than anyone in America. For example, data legislation more stringent than the US took a stride forward with the enactment of the GDPR, and hopefully more will be taken in the same direction with the ePrivacy Regulation. The EU has also dared to dish out sizable fines to tech-giants for not playing by the rules. A handful of American politicians have put digitalization at the forefront of their agendas. Gavin Newsom, governor in the hub of technological change itself, California, has been courageously vocal about the danger posed by technological unemployment. "We're going to get rolled over unless we get ahead of this," he told *The Guardian*. Many members of the US Congress consider Facebook, Google, and Amazon rulebreakers.[333]

The left has generally refrained from mentioning the inherent inequality of the digital economy. Sociologist Amitai Etzioni has adroitly noted how the left stays Keynesian on economic policy, but when it comes to new technology, they shift into *laissez-faire*.[334] Conservative parties, for their part, filibuster every kind of social reform, but then welcome all new tech regardless of its effect on society.

Many might call tech-criticism unconstructive due to it not fitting well into the game of traditional statecraft. There are, however, numerous platforms on which politicians with ration and a regard for well-being could stand. For example, there is a large margin to ease off the compulsion to digitize schools and government services. A heavier hand is needed over tech-giants that undermine both society and citizens, and to preempt them from evading taxation. Meaningful decisions that involve people should not be outsourced to the magic of algorithms unless the principles of their underlying code are fully disclosed. Instead of leaping headlong into each new technology, we should form a politic that fortifies jobs and cherishes happiness. The only responsible stance in UN discussions about autonomous weaponry is an outright ban, but the most pivotal decision of all is prohibiting anyone from developing AI that is more intelligent than us. Each nation has a monopoly on force to prevent such a crime against humanity, and if necessary, it should be used.

The Cul-de-sac of Progress

There are also those for whom the slowing and stopping of advancement is not enough, and who insist on savagely rewinding back the clocks. Famous for his bomb attacks and his literary critique of tech, Theodore Kaczynski writes that modern technology forms a unified system in which favorable developments which are encouraged cannot be distinguished from unfavorable ones which should be stopped. The system runs on the logic of unending expansion, the energy for which comes from the power struggle between people, and where new technologies are always befriended as helping. To Kaczynski, the inevitable technopalypse can only be avoided in one way: trash the modern technological society and go back to hunter-gatherer.[335]

Going with this alternative would no doubt lead to widespread human suffering. We have built everything on top of modern technology; society's main engines already spin according to the digital order. Should the newest techniques be suddenly abandoned, there would be no running water, buildings would no longer be heated, and hospitals would be forced to close their doors (the sun, though, would keep on shining). A state of disorder would prevail. It is clear that society's infrastructure can be designed without any leading-edge tech, but the political will necessary to make this transition is not forthcoming. A return to the developmental level of hunter-gatherer would cause a majority of the human population to succumb to starvation. Slowing or stopping technological development instead would not invite any catastrophe, despite those at the top of the food chain wanting to believe this.

As drenched in cliché as it sounds, the revolution that we

need must take place in the mind. We must reclaim the power to determine our own fate without prostrating ourselves before our digital devices. We must remember that reacting to technology's stimulus does not fill our day with what we really want. We must understand that society and culture can be constructed with things other than technology, and we should note how the tech god that we worship does not contain anything natural or intrinsic to us. Only recently have societies set their sights on maximizing the gain and efficiency of Bentham-Taylorism; the compulsive urge to advance, the obsession with "going forward," was not the mandate that most of history has lived by; a pervasive quest for profit and a surrender to market discretion were not the most celebrated social modii before the 1800s.

A proclivity to compete for status and fight over resources is, unfortunately, in our nature. These forces were already at work in primitive man and are today powering the economic and politico-military battlespaces, which are themselves busily driving tech development into absurdity. These inexorable components of the human being have surfaced at various points in history in various ways. Violent competition among males both within and between groups was once the norm. Still in the 1800s, America's first Treasury Secretary, Alexander Hamilton, and Vice President Aaron Burr resolved their disagreement with a pistol duel – killing Hamilton.[336] Today, such a bold and storied habit appears quackish.

Being endowed with the power of our digital wizardry puts us in the path of a long list of new issues, an eventual understanding of which will perhaps lead scientists, CEOs, and legislators into a new awareness of competition's limits: which crusade is necessary, and which should be bypassed as inadmissible. If it does not, the rogue mill of technology could grind everything to bits.

The Logical End

In addition to Alan Turing and John von Neumann, another essential ingredient in the birth of the computer was mathematician and logician Kurt Gödel. He was a man of small stature that fled Nazi Austria in the early part of WWII, and whose breakthroughs in the study of logic laid the foundation for computer programming.

Gödel was a paranoiac throughout his life, refusing even in old age to consume food made by anyone other than his wife, Adele. He was gripped by a fear that someone would poison him. During Adele's extended hospitalization in 1977, Gödel ate almost nothing at all, but his colleague attested to how malnutrition did not make Gödel's mind any less incisive.[337]

Months of this, however, left the 170cm Gödel with a weight of not even 30 kilograms. After Adele returned home from the hospital, she persuaded her weakening husband to seek admission to Princeton. Acute care was administered there to the thinning Gödel, but the man was a logician in every way. His interpretation of his health insurance contract departed from his doctors' – whose legalese was undoubtedly less developed than his – and he informed them that the procedure they offered was not covered under his policy. Gödel died in the hospital curled up in the fetal position on Saturday, January 14, 1978.[338]

The life and death of Gödel is rife with paradox, including one of his most famous scientific breakthroughs itself – the incompleteness theorems. These, in short, demonstrate that even in the realm of mathematics some things are beyond logical proof.[339] Gödel understood this in his work, but somehow not on his deathbed.

The almighty Machine rules our lives now more than 4 decades after Gödel's last breath, clacking mechanically on in its inanimate vitality, and if we do nothing, it will apply the principles of its perfect logic to the gloomy end.

Endnotes

1 Thoreau 2010, 46.

2 Nisbet 1980, 320.

3 Kurzweil 2005, 4.

4 Reuters 2013; Newlon 2013.

5 Kross & al. 2013; Tromholt 2016.

6 Frey & Osborne 2013; Pajarinen & Rouvinen 2014. See also: Brynjolfsson & McAfee 2014.

7 Carr 2010, 132.

8 Flynn & Shayer 2018; The Economist 2000; The Economist 2017d; OECD 2018a.

9 Bostrom 2014; Medeiros 2017; Dowd 2017.

10 Books were confiscated also in Brave New World, because "history is bunk." Huxley 1978, 46, 62; Orwell 2014.

11 Morris 2010; Vahvanen 2011.

12 Bury 1955, 93.

13 Morris 2010, Ch. 3.

14 Mokyr 1990, 183–190.

15 Lawler 2017.

16 Toffler 1972, 398–402; Citation on p. 398. Neither the Finnish Parliament's Committee for The Future nor the US Congressional Office of Technology Assessment (1972-1995) have yet done this.

17 Johnson 1998.

18 McLuhan 2003, Ch. 1; Carr 2010, 115–143.

19 Kaczynski 2010, 208.

20 Bird & Sherwin 2009, 319.

21 Easterlin 2001, 465.

22 Bury 1955, 7–20; Wright 1997, 2–3. Some researchers actually state that the idea of progress was already budding in classical antiquity. For example, Seneca is believed to have furthered some of the first conceptualizations of progress.

See: Nisbet 1980, 10–46; Burkert 1997, 19–46.

23 Bury 1955, 17.

24 Mokyr 1990, 201–208.

25 Wright 1997, 7.

26 Wright 1997, 6.

27 Nisbet 1980, 179–186.

28 Bury 1955, 222–223

29 Mokyr 1990, 31–56.

30 Bury 1955, 165–168.

31 Gurven & Kaplan 2007; Clark 2007, 2, 40–70, 91–111.

32 Clark 2012; The Economist 2014.

33 Berg 1980, 9, 12.

34 Brontë 1988, 41; Hobsbawm 1952; Mokyr 1992.

35 Clark 2007, 19–39. Malthus did not find population growth to be such an inevitable roadblock to progress as is often asserted. See: Nisbet 1980, 7, 178, 216–220.

36 Diamond 2004, 261.

37 Hobsbawm 1999, 659; Rees 2004, 13; Diamond 2004, 266–267.

38 Diamond 2004, 262–263.

39 Diamond 2004, 261.

40 Mokyr 1990, 55.

41 Mokyr 1990, 167–170; Hobsbawm 1999, 656–657.

42 Hobsbawm 1999, 657.

43 Toffler 1981, 98–103.

44 Thiel 2011. Flying cars will already soon be available. See: Smith 2017.

45 Raittila 2004; The Economist 2013.

46 Rousseau 2009, 74; Hobbes 1996, 89. Rousseau's ideal group was not that of primitive man, but rather somewhere between it and modern society. See: Bury 1955, 179–180.

47 Nisbet 1980, 180. The same kind of channeling of selfishness and evil into the greater good – with the help of established institutions – was examined by Niccolò Machiavelli in the

1500s before Turgot. See Fukuyama 1992, 234.

48 According to these statistics, between Democracies and Authoritocracies was a cross between them, the Anacrocies. See: Roser 2018. Freedom House defines democracy in slightly stricter terms than Roser's Polity IV Score. According to Freedom House, less than half of all countries were democracies in 2018. See: Freedom House 2018.

49 Pinker 2012.

50 Bauman 1989.

51 NASA 2018.

52 The Economist 2016a; Ellen MacArthur Foundation 2016; Boakes 2018.

53 Bawden 2016; Walsh 2013.

54 Wright 2004, 38–40. Walter Von Krämer and Daniel B. O'Leary raised the concept of the progress trap in their work before Wright. Historian Ian Morris alludes to it when writing about the paradoxes of development.Wright 2004, 77–79.

56 Rees 2004.

57 Easterlin 1974; Clark et al. 2018. For a critique of the Easterlin Paradox, see: Stevenson & Wolfers 2008. Easterlin's response: Easterlin & al. 2010.

58 $75,000 (~65,000 EUR) is the threshold beyond which more money does not increase happiness. When asked to rate their own life and its success – instead of happiness, respondents did not indicate such a "happiness saturation point." Kahneman & Deaton 2010.

59 For example, see: Clark 2007, Clark et al. 2018, 70–80, 376; Kahneman et al. 2006. Research suggests an emphasis on mental health services could be several times cheaper at increasing happiness than efforts to reduce poverty. See: Pilling 2018, 253.

60 Layard 2005, 32–35; Pantsu 2018; Wilkinson & Pickett 2011, 36.

61 Pilke 2013; Enbuske 2014; Pääkkönen & Hanif 2011.
62 Gourevitch 1997, 203; Talsi 2014.
63 Morris 2010, 537.
64 Linkola 1989, 187.
65 Eliot 2015, 153.
66 Carr 2010, 107, 115–143.
67 Burke 2016.
68 Stephens-Davidowitz 2017, 220.
69 Review articles about the effects of digital devices on intelligent functioning in general: Wilmer et al. 2017; Loh & Kanai 2016.
70 Leiva et al. 2012.
71 Sana et al. 2013.
72 Adler & Benbunan-Fich 2013.
73 Weisberg 2016; Stothart et al. 2015; Thornton et al. 2014.
74 Interview with Mona Moisala. 23 Feb 2018.
75 Moisala 2017. In Moisala's dissertation (comprised of three different studies), it was observed that playing computer games can improve short-term memory and reaction speed. Ophir et al. 2009 reached the same sort of conclusions as Moisala regarding concentration issues in connection with multitasking. See also: Loh & Kanai 2014; Uncapher et al. 2016. Those who regularly multitask could show a greater ability to seamlessly switch from one task to another: Alzahabi & Becker 2013.
76 Interview with Mona Moisala. 23 Feb 2018.
77 Hadar et al. 2015. Regarding the degradation of impulse control, see also: Wilmer & Chein 2016.
78 Vinter 2012.
79 Raeste 2017.
80 Karpicke & Blunt 2011. See also: Slamecka & Graf 1978.
81 Sparrow et al. 2011.
82 Henkel 2014.
83 Carr 2010, 123–124.

84 Carr 2014, 1.

85 Carr 2014, 43–45.

86 Ibid., 70–71; Hansen 2013; California Maritime Academy 2018. There is also evidence that excessive dependence on GPS can weaken memory. Experiments on rats by Nobel prize-winning Norwegian neuroscientists May-Britt and Edvard Moser indicate that the same areas of the brain that control special orientation also control memory. See: Abbott 2014; Gorman 2013.

87 Turkle 2009, 55–56, 70; Alberdi et al. 2008.

88 Kellaway 2016.

89 Björksten 2016.

90 Rosen et al. 2013; Junco & Cotten 2012; Lepp et al. 2014; Karpinski et al. 2013; Jacobsen & Forste 2011.

91 Beland & Murphy 2014, 18; Hirvonen 2018. In summer 2018, Pasi Sahlberg, an expert on Finland's internationally acclaimed educational system, submitted that smartphones are likely one factor behind the drop in PISA scores in many developed nations. See: Honka 2018.

92 Anderson 2015. An update came into effect with the 2016 SAT tests making scores no longer comparable to earlier ones. For more about the PISA's problems in mapping development, see: Kivinen & Hedman 2017.

93 Wolf & Barzillai 2009.

94 Flynn & Shayer 2018. Cuddy 2017. See also: Bratsberg & Rogeberg 2018.

95 Moody 2018.

96 Aamulehti 2016.

97 OECD 2015, 3–4, 53–54.

98 Ibid., 3.

99 Ibid., 3; Prime Minister's Office 2015. There is hope with respect to new tech in the classroom. Study applications that answer to individual needs could improve scores. This is not unequivocally verified by research, the majority of

which was done in India, whose classroom environment is much different than in Western countries. See: The Economist 2017a.

100 Bilton 2014; Weller 2017; Yates 2017.

101 Bilton 2014.

102 Fessler 2017; Bauman 1989, 6 –7; Borowski 2008.

103 McLuhan 1964, Ch. 1.

104 Greenfield 2014, 2.

105 Platon 1925.

106 Thoreau 2010, 46.

107 Pääkkönen & Hanifi 2011, 38.

108 Carr 2010, 136–137; Haile 2014.

109 Carr 2010, 138.

110 Schonfeld 2009.

111 Greenfield, 2014, 21–22.

112 Van der Weyde 1925.

113 Parker 2004.

114 McKew 2018; RoBhat Labs 2017.

115 McKew 2018.

116 Sunstein 2017, 1–3.

117 2018.Ibid., 6–7; The Economist 2018g.

118 Sunstein 2017, 7.

119 Singal 2018.

120 Sunstein 2017, 10.

121 Levitsky & Ziblatt 2018, 8–9.

122 Epstein & Robertson 2015; Lazareva 2018. On the other hand, the internet has created networks that challenge centralized information – for example, right-wing websites that often border on, or blatantly propound, fake news. See: Cadwalladr 2016.

123 Evans 2008.

124 Wong 2017.

125 The Economist 2016g; Subramanian 2017.

126 Isaac & Wakabayashi 2017; Green & Issenberg 2016.

127 Perera 2017; The Economist 2018f.

128 Specia & Mozur 2017; McLaughlin 2018; Harris 2018; Pasternack 2017.

129 For fake news produced by the USSR in the 1980s, see: Qiu 2017.

130 Carey 2018.

131 Howard et al. 2017; Blake 2018.

132 Levin et al. 2016; Poutanen 2018; Wong 2017.

133 Orwell 2014, 13.

134 Palihapitiya 2017.

135 Stephens-Dawidowitz 2017, 4, 19, 123–140. With the Google Trends feature, one can observe how the standing of various search words has changed over the preceding year.

136 Ibid., 50–52, 121.

137 Interview with Juha Nurmi 7 Apr 2018; Hackett 2015; Ward 2017; Greenwald et al. 2013.

138 Kosinski et al. 2013.

139 Wu et al. 2015.

140 Ibid.

141 Ibid.

142 Grassegger & Krogerus 2017; Timberg et al. 2018; Weaver 2018. Information was also collected from Twitter. See: Fingas 2018.

143 Hern 2018; Grassegger & Krogerus 2017. Presentation on the limits of Cambridge Analytica's methods: Chen 2018.

144 Channel 4 News 2018.

145 Allen & Abbruzzese 2018.

146 McCormak 2016; Matz et al. 2017.

147 Tufekci 2012.

148 Ibid.; Dwoskin & Romm 2018.

149 Allen & Parker 2016; O'Brien 2017.

150 Lewis 2018; Dance et al. 2018.

151 Madrigal 2018b; Tufekci 2018.

152 Kramer et al. 2014.
153 Bond et al. 2012.
154 Zittrain 2014; Bowles & Thielman 2016.
155 Facebook 2016; Garcia-Martinez 2018; Green & Issenberg 2016. In 1996, Hillary Clinton spoke about "superpredators" prowling the streets of America, but did not refer to blacks. See: Clinton 1996.
156 Epstein & Robertson 2015.
157 European Commission 2017. Google received an even bigger fine in summer 2018 (4.3 Billion EUR) for recommending its own products on Android. See: BBC 2018.
158 Epstein & Robertson 2015.
159 Duhigg 2012.
160 O'Neil 2016, 70.
161 Ibid., 72.
162 The National Non-Discrimination and Equality Tribunal 2018; Interview with Financial Supervisory Authority risk analyst Torsten Groschup 30 Aug 2018.
163 The Economist 2016f.
164 O'Neil 2016, 105–122.
165 Ibid., 3–11, 130–134; Stephens-Davidowitz, 261–262.
166 Wu et al. 2015; Interview with Zoltan Istvan 9 May 2015.
167 Auvinen-Lintunen et al. 2015; Stephens-Davidowitz 2017, 266–269.
168 Gorner 2013; O'Neil 2016, 102.
169 Interview with Sari Sarani 10 Apr 2018.
170 Ibid.
171 Electronic Privacy Information Center 2018. According to one study, algorithms are at the level of the average human in predicting repeat offenses, but not as capable as experts. See: The Economist 2018c.
172 FBI 2014; Larson et al. 2016.
173 Stephens-Davidowitz 2017, 262.
174 Witte 2013; Schneier 2014.

175 The Physics ArXiv Blog 2014; Laine 2018; Metz & Singer 2018.
176 Wu & Zhang 2016.
177 The Economist 2017b.
178 Andrade 2014; Haapaluoma-Höglund 2017.
179 Matikainen 2018.
180 Hvistendahl 2017; Greenfield 2018.
181 Hvistendahl 2017; Greenfield 2018.
182 Penney 2017.
183 Orwell 2014.
184 Kirn 2010.
185 I wrote about these in the weekend edition of the Helsingin Sanomat: Vahvanen 2017.
186 Wikiquote 2018.
187 Interview with Juha Nurmi 7 Apr 2018.
188 Interview with Erka Koivunen 13 Apr 2018.
189 Streitfeld et al. 2018; Schneier 2018; Interview with Reijo Aarnio 29 May 2018; Interview with Erka Koivunen 13 Apr 2018; Singer 2018; Kang 2018.
190 O'Neil, 2016, 208–212; Wachter et al. 2016.
191 Tufekci 2012.
192 Hirvonen 2015.
193 The Economist 2017c. Additionally, the anonymity of individual DNA can be deciphered by comparing it with its closest matches – i.e., family members. This method is utilized in the US in identification efforts in cases of serious crime. See: Murphy 2018.
194 Dwoskin et al. 2018b; Metzger 2016; Burgess 2017.
195 Mustonen 2018.
196 Interview with Erka Koivunen 13 Apr 2018.
197 Nietzsche 1995, 17–18.
198 Goethe 1981a, 20.
199 I am well aware of the fact that many take the use of the word "spirit" to be uncalled for. Despite its dated nature

and religious connotation, I find it in this circumstance to be simultaneously broader and more pertinent than, for instance, "psyche."

200 Theodore Kaczynski addresses this with the concept of Power Process – meaning a person should have a feeling of having power by which he can bring his goals to fruition. See: Kaczynski 1995.

201 I first wrote about technological society's enjoyment-by-keystroke in the Helsingin Sanomat. See: Vahvanen 2014.

202 I first wrote about the culture of insatiability in my Emotional Overload essay for the Sunnuntaisuomalainen 5 Jan 2014.

203 Simmel 2005.

204 Toffler 1972, 75–92. Significant social media use is linked in research to relationship difficulties and a quicker decision to part ways. See: Clayton et al. 2013; Clayton 2014.

205 Turkle 2012, 1.

206 Ducharme 2018; Greenfield 2014, 10.

207 Karl Marx writes, "All our invention and progress seem to result in endowing material forces with intellectual life, and in stultifying human life into a material force." See Carr 2014, 24. For the effects of mobile devices on the comprehension of emotional expression, see: Uhls et al. 2014.

208 Turkle 2015; Smith 2015; Saari 2017.

209 Konrath et al. 2011.

210 Turkle 2012, 3–4.

211 Kontula 2015.

212 Rubin 2018; Levy 2009, 22.

213 Weizenbaum 1984, 2–8, 190–191.

214 Cf. Turkle 2012, 282.

215 Ogburn 1922, 200–280. The same kind of view of the imbalance between technological and cultural change was already evident in the work of sociologist Thorstein Veblen

20 years prior.

216 Fukuyama 1992. Thumos is often spelled Thymos.

217 Nietzsche 1995, 17–19. Fukuyama 1992, 368–382.

218 Twenge 2017.

219 Ibid.

220 Ibid.

221 Kross et al. 2013; Shakya & Christakis 2017; Levin 2017.

222 Tromholt 2016; Verduyn et al. 2015; Campbell 2017.

223 Verduyn et al. 2017.

224 Kunttu et al. 2016. Researchers are still not fully unanimous about whether psychiatric problems have gone up among children, adolescents, and adults. One should always ask: can the apparent rise in diagnoses be in part the result of a more sensitive search for treatment? This would not, however, entirely explain the change. See, for example: Collishaw 2014; Olfson et al. 2014.

225 OECD 2017a; OECD 2017b; The Economist 2018a; The Economist 2018b.

226 Vänttinen 2016.

227 OECD 2017b.

228 Kaczynski 1995.

229 The Economist 2018b.

230 Lewis 2017b; Dilts 2018; Reuters 2017. Valinsky 2018.

231 Ford 2015, 32.

232 Mokyr 1990, 155.

233 Casti & DePauli 2001, 152–155; Brynjolfsson & McAfee 2011, 12.

234 Brynjolfsson & McAfee 2011, 13–14.

235 Morris 2010, 593.

236 The Economist 2016b; Bohn 2018.

237 Miller 2012, 4; Markoff 2011; Russell 2017; Metz 2017a.

238 Barrett 2017. Moore himself predicted in 1995 that the law that bears his name would come to an end in 2005; In 2005, his prediction was 2015–2025, and in 2015, it was 2025. See:

The Economist 2015.

239 Bell 2016.

240 AFP 2018; Lohr 2017; Vincent 2017b; Junkkari & Teittinen 2016; Liimatainen 2016.

241 Kaplan 2015.

242 Frey & Osborne 2013; Pajarinen & Rouvinen 2014. Ari Ojapelto has long spoken and written about the threat of technological unemployment in Finland.

243 Acemoglu & Restrepo 2017. McKinsey 2017a; McKinsey 2017b.

244 Mokyr 1990 256–257; Mokyr 1992, 331–332.

245 Mokyr 1992, 336–337.

246 The Economist 2016a; The Economist 2016c; Ford 2015, 30–31; Acemoglu & Restrepo 2017, 1.

247 Autor 6–7.

248 Ibid., 5, 7.

249 Tiainen 2014; Findicator 2018; OECD 2018b; Federal Reserve Bank of St Louis 2018.

250 McKinsey Global Institute 2015.

251 The word "robot" is known to have first been used by Czech author Karel Čapek in 1921 in his play RUR (Rossumovi Univerzální Roboti). The word comes from the Czech "robota," which means forced labor.

252 Cf. Collins 2014, 50–90.

253 Paul Krugman writes that the modeling of economists that romanticized Capitalism "led most economists to ignore all the things that can go wrong." See: Krugman 2009. Thomas Piketty writes that over the 1900s, "economists' overly developed taste for apocalyptic predictions gave way to a similarly excessive fondness for fairy tales, or at any rate happy endings." Piketty 2016, 22.

254 Bynjolfsson & McAfee 2014, 126–128; McCormick 2018; Aslam 2018.

255 Douglas 2018; Ting 2016; Statista 2018; Uber 2018; Trefis

Team 2018.

256 McKinsey 2017b, 28.

257 Smith & Anderson 2014; Junkkari & Liimatainen 2016; Vincent 2018; Pulkka 2017.

258 Carr 2014, 14–15.

259 Brooks 2013.

260 Young 2012; Blakely et al. 2003; Lynge 1997.

261 ILO & OECD 2015.

262 Brynjolfsson & McAfee 2014, 147–162. This logic still cannot be applied to every sector. Masseuses have to limit the number of massages they give and dentists must limit the number of patients they treat.

263 By this metric, inequality in the US is greater than it was at the start of the 1900s, when the structure of American society was in many ways fundamentally more unequal. The spread of democracy, education, and progressive social policy kept the wealthy class in check until the end of the 1900s, after which inequality expanded significantly. Brynjolfsson & McAfee 2014, 164; Piketty 2016, 285; Sommeiller & Price 2018.

264 Darvas 2018; Tilastokeskus 2017.

265 Mitrunen 2013; Ford 2015, 49–51; Michaels et al. 2014; Kopf 2017; Cowen 2013.

266 Ford 2015, 53–58; Interview with Risto Heiskala 5 Jul 2018; Summers 2018.

267 The Economist 2018d; Tiku 2017.

268 The Economist 2018d; Brännare 2017; Luotonen & Lehtovaara 2018. The markets are even more built up in China. Almost everyone uses WeChat, by which one can manage messaging, calls, net searches, games, and payments all in one. Didi, China's Uber, owns 90 percent of the market. Alibaba is in control of 60 per cent of the online shopping share. See: The Economist 2016d; Kwok 2018.

269 Interview with Martin Ford 11 Jun 2013. Perkiö & Pulkka

2018, 412–413. Karvala 2018. One very seldom discussed alternative to a Basic Income is a State Job Guarantee, where the government guarantees work for those who are not employed in the market. In addition to ensuring individual income, this would potentially decrease the workforce shortage in the public sector. See: Perkiö & Pulkka, 419.

270 Berg 1980, 243. Delaney 2017; Price 2017; Yoon 2017; Interview with Mika Maliranta 25 Jul 2018.

271 Koski & Husso 2018.

272 Mokyr 1992, 329.

273 Pilling 2018, 252; Young 2012; Toffler 1972, 298–300.

274 The creator of the GDP metric, the late Simon Kuznets, was himself opposed to his indicator being used to gage well-being. Pilling 2018, 3, 19–44.

275 Pilling 2018, 241. For example, the health of mind and body, one's spouse, and having a job are, in addition to salary, among the most important factors in enhancing happiness. See: Clark et al. 2018, 36–41.

276 Polanyi 2008.

277 Ibid.; Goethe 1981b.

278 Mascarenhas 2017.

279 Vincent 2017a.

280 Butler 1910, Ch. 23.

281 Ibid.

282 Mascarenhas 2017.

283 McFarland 2014; Dowd 2017.

284 Bostrom 2014, 18–21.

285 Headrick 2010, 101–112, 257–292. Mao's quote: Morris 2010, 431. Advanced medical technology greatly assisted global conquest. It was used to fight disease in Africa, for instance.

286 Wright 2004, 7.

287 Bird & Sherwin 2009, 323. During the war, Oppenheimer

argued that the employment of the atomic bomb on Japan would showcase the preposterousness of the use of such weapons and therefore make future war impossible. "Lots of boys not grown up yet will owe their life to it." Ibid., 302, 309.

288 Dyson 2013, ix, 329.

289 Vinge 1993.

290 Ibid.

291 Ibid.; Bostrom, 62–65.

292 Asimov 1942.

293 Bostrom 138–142. See also: Yudkowsky 2008.

294 Vincent 2016.

295 Bostrom 2014, 128–131.

296 Metz 2015; Metz 2017b.

297 Schneider 2016.

298 Pearl 2018; Hartnett 2018.

299 Regarding the possible emergence of AI consciousness, see, for example: McDermott 2007.

300 Interview with Nick Bostrom 21 Oct 2014.

301 Kurzweil 2005, 21–34.

302 Harris 2017; Roy 2014.

303 Berman 2011; Vlahos 2017.

304 Popper 2014; Roy 2014.

305 Vincent 2017a; Sample 2018; Metz 2018.

306 Singer 2009.

307 Warrick 2017.

308 Carpenter 2013; Madrigal 2018a; Lamothe 2015; The Economist 2018e; Haltiwanger 2017; Russell 2015; Scharre 2018. Very basic and nearly autonomous weapon systems have been around already for longer. The Aegis system of the US naval fleet has long been capable of shooting down enemy fighters and missiles on its own in the absence of human intervention. In 1988, the system shot down an Iranian passenger aircraft in the Persian Gulf by mistake.

Despite doubts, the vessel crew did not interfere with its operation. All 290 people aboard the aircraft lost their lives. See: Singer 2009.

309 US Department of Defense 2011, notably 43–51; US Department of Defense 2012; Carpenter 2013; US Army 2017.

310 Singer 2009.

311 Human Rights Watch 2016; US Government 2018.

312 Solon 2016; Russell 2015.

313 Stop Killer Robots Campaign 2015; Shane et al. 2018; Metz 2018.

314 Busby 2018; Interview with Pekka Appelqvist 11 Jun 2018. A possible treaty would apply to offensive autonomous weapons, not defensive – which include automatic air defense systems and missile defense systems.

315 Broscious 1999, 24–38. Citation on p. 36.

316 Rees 2004, 73–88.

317 Rees 2004, 26.

318 Butler 1863.

319 Salvadori 1968, 242.

320 Kharpal 2017.

321 Freud 2002, 26.

322 Drexler 1986; Jha 2011.

323 Brown 2012; Walters 2014; BBC 2007.

324 Fukuyama 2002; Keen 2013, 57; Lawler 2017.

325 Turing's official cause of death was ruled as suicide. Some have refuted this, stating that he was murdered. See Leavitt 2006, 275–280.

326 Pinker 2012, 571–670.

327 Berg 1980, 253–290.

328 Ibid., 10.

329 Lewis 2017b; Interview with Juha Nurmi 7 Apr 2018.

330 Collins 2014, 76–89; Burke 2017; Chrisafis 2016; Hellier 2015.

331 La Casemate 2017; Earth First! 2017.

332 Beckhusen 2013; Harding 2010; Interview with Brian Nussbaum 10 Feb 2015.

333 Lewis 2017a.

334 Toffler 1972, 398; Kempas 2018.

335 In Kaczynski's opinion, such a thing would be even more justified than the campaign against Nazi Germany in WWII, because technology can do humanity in forever (mentally as well as physically), and transcends the span of any one totalitarian regime. See: Kaczynski 2010, 243; Kaczynski 2016.

336 Pinker 2012, 22.

337 Goldstein 2005, 224, 249–250.

338 Menger 1994, 232; Goldstein 2005, 251.

339 Goldstein 2005, 165.

Bibliography

Aamulehti: Editorial: "Pisa-tulokset on syytä ottaa vakavasti, mutta niihin ei kannata hirttäytyä," 8 Dec 2016. Available at: https://www.aamulehti.fi/paakirjoitukset/paakirjoitus-pisa-tulokset-on-syyta-ottaa-vakavasti-mutta-niihin-ei-kannata-hirttaytya-24128334/

Abbott, Alison: "Neuroscience: Brains of Norway," *Nature*, 6 Oct 2014. Available at: http://www.nature.com/news/neuroscience-brains-of-norway-1.16079

Acemoglu, Daron & Restrepo, Pascual: Robots and Jobs: Evidence from US Labor Markets. National Bureau of Economic Research, Working paper 23285, March 2017. Available at: http://www.nber.org/papers/w23285.pdf

Adler, Rachel F. & Benbunan-Fich, Raquel: Self-interruptions in discretionary multitasking. *Computers in Human Behavior*, July 2013, 1441–1449. Available at: https://pdfs.semanticscholar.org/5b5e/4d6fe36dfc7795c6f40388dd72e7f5ef3735.pdf

AFP: "Computer learns to detect skin cancer more accurately than doctors," *The Guardian*, 29 May 2018. Available at: https://www.theguardian.com/society/2018/may/29/skin-cancer-computer-learns-to-detect-skin-cancer-more-accurately-than-a-doctor

Alberdi, E.; Povyakalo, A. A.; Strigini, L.; Ayton, P. & Given-Wilson, R. M.: CAD in mammography: lesion-level versus case-level analysis of the effects of prompts on human decisions. *International Journal of Computer Assisted Radiology and Surgery*, June 2008, 115–122. Available at: http://openaccess.city.ac.uk/1583/1/JCARS08.pdf

Allen, Jonathan & Abbruzzese, Jason: "Cambridge Analytica's effectiveness called into question despite alleged Facebook data harvesting," NBC News, 20 Mar 2018. Available at: https://www.nbcnews.com/politics/politics-news/cambridge-analytica-s-effectiveness-called-question-despite-alleged-facebook-data-n858256

Allen, Kate & Parker, George: "On the In-side: the campaign to keep Britain in the EU," *The Financial Times*, 13 Jun 2016. Available at: https://www.ft.com/content/6ab870d6-2f01-11e6-bf8d-26294ad519fc

Alzahabi, R. & Becker, M.W.: The association between media multitasking, task-switching, and dual-task performance. *Journal of Experimental Psychology: Human Perception and Performance*, October 2013, 1485–1495. Available at: https://www.ncbi.nlm.nih.gov/pubmed/23398256

Anderson, Nick: "SAT scores at lowest level in 10 years, fueling worries about high schools," *The Washington Post* 3 Mar 2015. Available at: https://www.washingtonpost.com/local/education/sat-scores-at-lowest-level-in-10-years-fueling-worries-about-high-schools/2015/09/02/6b73ec66-5190-11e5-9812-92d5948a40f8_story.html?utm_term=.cdbe1d807cea

Andrade, Norberto: "Computers are Getting Better than Humans at Facial Recognition," *The Atlantic*, 9 Jun 2014. Available at: https://www.theatlantic.com/technology/archive/2014/06/bad-news-computers-are-getting-better-than-we-are-at-facial-recognition/372377/

Asimov, Isaac: *Runaround* (1942). Available at: http://web.williams.edu/Mathematics/sjmiller/public_html/105Sp10/handouts/Runaround.html

Aslam, Salman: "Instagram by the Numbers: Stats, Demographics & Fun Facts," Omnicore, 1 Jan 2018. Available at: https://www.omnicoreagency.com/instagram-statistics/

Autor, David: Why Are There Still So Many Jobs? The History and Future of Workplace Automation. *Journal of Economic Perspectives*, Summer 2015, 3–30. Available at: https://economics.mit.edu/files/11563

Auvinen-Lintunen, Laura; Häkkänen-Nyholm, Helinä; Ilonen,

Tuula & Tikkanen, Roope: Sex Differences in Homicidal Fantasies among Finnish University Students. *Psychology*, 14 Jan 2015, 39–47. Available at: http://file.scirp.org/pdf/PSYCH_2015011416291501.pdf

Barrett, Brian: "An IBM breakthrough ensures silicon will keep shrinking," *Wired*, 6 May 2017. Available at: https://www.wired.com/2017/06/ibm-silicon-nanosheets-transistors/

Bauman, Zygmunt: *Modernity and the Holocaust* (Cornell University Press: 1989).

Bawden, Tom: "Global warming: Data centres to consume three times as much energy in the next decade, experts warn," *The Independent*, 23 Jan 2016. Available at: http://www.independent.co.uk/environment/global-warming-data-centres-to-consume-three-times-as-much-energy-in-next-decade-experts-warn-a6830086.html#gallery

BBC: "Leaking drains 'caused outbreak,'" 7 Sep 2007. Available at: http://news.bbc.co.uk/2/hi/uk/6982709.stm

BBC: "Google hit with €4.3bn Android fine from EU," 18 Jul 2018. Available at: https://www.bbc.com/news/technology-44858238

Beckhusen, Robert: "In Manifesto, Mexican Eco-Terrorists Declare War on Nanotechnology," *Wired*, 3 Dec 2013. Available at: https://www.wired.com/2013/03/mexican-ecoterrorism/

Beland, L.; Murphy, R.J.: Ill Communication: Technology, Distraction & Student Performance. London School of Economics and Political Science, 2014. Available at: http://cep.lse.ac.uk/pubs/download/dp1350.pdf

Bell, Daniel: *The Cultural Contradictions of Capitalism* (Basic Books: 1976).

Bell, Lee: "Machine learning versus AI: what's the difference?," *Wired*, 1 Dec 2016. Available at: http://www.wired.co.uk/article/machine-learning-ai-explained

Berg, Maxine: *The machinery question and the making of political economy 1815–1848* (Cambridge University Press: 1980).

Berman, John: "Futurist Ray Kurzweil Says He Can Bring His Dead Father Back to Life Through a Computer Avatar," ABC News, 9 Aug 2011. Available at: https://abcnews.go.com/Technology/futurist-ray-kurzweil-bring-dead-father-back-life/story?id=14267712

Bilton, Nick: "Steve Jobs was a Low-Tech Parent," *The New York Times*, 10 Sep 2014. Available at: https://www.nytimes.com/2014/09/11/fashion/steve-jobs-apple-was-a-low-tech-parent.html

Bird, Kai & Sherwin, Martin J.: *American Prometheus: The Triumph and Tragedy of J. Robert Oppenheimer* (Atlantic Books 2009; orig. 2005).

Björksten, Tuomo: "Uudet Pisa-tulokset: Suomen tytöt maailman toiseksi parhaita, vaikka ei edes huvita," Yle Uutiset 6 Dec 2016. Available at: https://yle.fi/uutiset/3-9336636

Blake, Aaron: "A new study suggests fake news might have won Donald Trump the 2016 election," *The Washington Post* 3 Apr 2018. Available at: https://www.washingtonpost.com/news/the-fix/wp/2018/04/03/a-new-study-suggests-fake-news-might-have-won-donald-trump-the-2016-election/?utm_term=.12fead2c745d

Blakely, T. A.; Collings, S. C. D.; Atkinson, J.: Unemployment and Suicide. Evidence for a Causal Association? *Epidemology & Community Health*, August 2003, 594–600. Available at: http://jech.bmj.com/content/57/8/594.full

Boakes, Elizabeth: "Extinction is happening at 1,000 times the normal speed," Quartz, 9 Jul 2018. Available at: https://qz.com/1323792/extinction-is-happening-at-1000-times-the-normal-speed/

Bohn, Dieter: "What it's like to watch an IBM AI successfully debate humans," The Verge, 18 Jun 2018. Available at: https://www.theverge.com/2018/6/18/17477686/ibm-project-debater-ai

Bond, Robert M.; Fariss, Christopher J.; Jones, Jason J.; Kramer, Adam D.; Marlow, Cameron; Settle, Jaime E. & Fowler, James H.: A 61-Million-person experiment in social influence and political mobilization. *Nature*, 13 Sep 2012, 295–298.

Borowski, Tadeusz: *Kotimme Auschwitz*. Original Polish *Pozegnanie z Maria i inne*, 1946. Translated into Finnish by Martti Puukko (Like: 2008).

Bostrom, Nick: *Superintelligence: Paths, Dangers, Strategies* (Oxford University Press: 2014).

Bowles, Nellie & Thielman, Sam: "Facebook accused of censoring conservatives, report says," *The Guardian*, 9 May 2016. Available at: https://www.theguardian.com/technology/2016/may/09/facebook-newsfeed-censor-conservative-news

Brännare, Stina: "Jättiyhtiöt Google ja Facebook kahmaisevat yhä suuremman osuuden Suomen digimainonnasta," Yle Uutiset, 28 Jun 2017. Available at: https://yle.fi/uutiset/3-969159Bratsberg, Bernt & Rogeberg, Ole: Flynn effect and its reversal are both environmentally caused. *Proceedings of the National Academy of Sciences of the United States of America*. Published electronically 11 Jun 2018. Available at: https://doi.org/10.1073/pnas.1718793115

Brontë, Charlotte: *Shirley*. Translated into Finnish by Kaarina Ruohtula. (Karisto: 1988; orig. 1849).

Brooks, Arthur C.: "A Formula for Happiness," *The New York Times*, 14 Dec 2013. Available at: https://www.nytimes.com/2013/12/15/opinion/sunday/a-formula-for-happiness.html?pagewanted=1&bicmp=AD&_r=1&bicmet=1388638800000&bicmst=1385874000000&bicmlukp=WT.mc_id&smid=fb-nytimes&fblinkge0

Broscious, S. David.: Longing for International Control, Banking on American Superiority: Harry S. Truman's Approach to Nuclear Weapons. Teoksessa Gaddis, John Lewis; Gordon, Philip H.; May, Ernest R. & Rosenberg, Jonathan (ed.): *Cold War Statesmen Confront the Bomb* (Oxford University Press: 1999).

Brown, David: "One of two controversial 'bird flu' papers is published," *The Washington Post*, 2 May 2012. Available at: https://www.washingtonpost.com/national/health-science/one-of-two-controversial-bird-flu-papers-is-published/2012/05/02/gIQA51egxT_story.html?utm_term=.a4db2392eca0

Brynjolfsson, Erik & McAfee, Andrew: *Race Against the Machine: How the Digital Revolution is Accelerating Innovation, Driving Productivity, and Irreversibly Transforming Employment and the Economy* (Digital Frontier Press: 2011).

Brynjolfsson, Erik & McAfee, Andrew: *The Second Machine Age: Work, Progress, and Prosperity in a Time of Brilliant Technologies* (W. W. Norton: 2014).

Burgess, Matt: "Despite the controversy, DeepMind signs up another NHS trust to use its Streams app," *Wired*, 22 Jun 2017. Available at: http://www.wired.co.uk/article/deepmind-nhs-streams-deal

Burke, Claire: "Distracted in the Office? Blame Evolution," *The Guardian*, 1 Mar 2016. Available at: https://www.theguardian.com/careers/2016/mar/01/distracted-office-blame-evolution-workspace-design-focus

Burke, Jason: Violence erupts between taxi and Uber drivers in Johannesburg. *The Guardian* 8 Sep 2017. Available at: https://www.theguardian.com/world/2017/sep/08/violence-erupts-taxi-uber-drivers-johannesburg

Burkert, Walter: Impacts and Limits of the Idea of Progress in Antiquity, Burgen, Arnold; McLaughlin, Peter & Mittelstraß, Jürgen (ed.): *The Idea of Progress* (de Gruyter: 1997), 19–46.

Bury, J. B.: *The Idea of Progress – An Inquiry into Its Origin and Growth* (Dover Publications: 1955; orig. 1922).

Busby, Mattha: "Killer robots: pressure builds for ban as governments meet," *The Guardian*, 9 Apr 2018. Available at: https://www.theguardian.com/technology/2018/apr/09/killer-robots-pressure-builds-for-ban-as-governments-meet

Business Finland, Ministry of Transport and Communications,

Technology Industry and Internet Industry: *Digibarometri 2018*, 6 Jun 2018, Taloustieto Oy. Available at: http://www.digibarometri.fi

Butler, Samuel [Original author noted as Cellarius]: "Darwin Among the Machines," *The Press*, 13 Jun 1863. Available at: http://nzetc.victoria.ac.nz/tm/scholarly/tei-ButFir-t1-g1-t1-g1-t4-body.html

Butler, Samuel: *Erewhon* (A.C. Fifield: 1910; orig. 1872). Available at: http://www.gutenberg.org/files/1906/1906-h/1906-h.htm

Cadwalladr, Carole: "Google, democracy and the truth about internet search," *The Guardian*, 4 Dec 2016. Available at: https://www.theguardian.com/technology/2016/dec/04/google-democracy-truth-internet-search-facebook

Californian Maritime Academy: *Overreliance on Modern Navigation Aids: The Role of Traditional Methods of Navigation and Case Studies*. (No date in report, 29 Aug 2018 noted on webpage) Available at: HTTPS://WWW.CSUM.EDU/C/DOCUMENT_LIBRARY/GET_FILE?UUID=79616282-5BCB-4E15-9341-93E996779CA4&GROUPID=61938

Campbell, Denis: "Facebook and Twitter 'harm young people's mental health'" *The Guardian*, 19 May 2017. Available at: https://www.theguardian.com/society/2017/may/19/popular-social-media-sites-harm-young-peoples-mental-health

Carey, Benedict: "'Fake News': Wide Reach but Little Impact, Study Suggests," *The New York Times*, 2 Jan 2018. Available at: https://www.nytimes.com/2018/01/02/health/fake-news-conservative-liberal.html

Carpenter, Charli: "Beware the Killer Robots," *Foreign Affairs*, 3 Jul 2013. Available at: https://www.foreignaffairs.com/articles/united-states/2013-07-03/beware-killer-robots;

Carr, Nicholas: *The Shallows – How the Internet is changing the way we think, read and remember* (Atlantic Books: 2010).

Carr, Nicholas: *The Class Cage: How Our Computers are Changing*

Us (W. W. Norton & Company: 2014).

Casti, John L. & DePauli, Werner: *Kurt Gödel – Elämä ja matematiikka.* Original English *Kurt Gödel: A Life of Logic,* 2000. Translated into Finnish by Risto Vilkko (Art House: 2001).

Channel 4 News: "Cambridge Analytica Uncovered: Secret filming reveals election tricks," 19 Mar 2018. Available at: https://www.youtube.com/watch?v=mpbeOCKZFfQ ja https://www.youtube.com/watch?v=cy-9iciNF1A&t=281s

Chen, Angela & Potenza, Alessandra: "Cambridge Analytica's Facebook Data Abuse Shouldn't Get Credit for Trump," The Verge, 20 Mar 2018. Available at: https://www.theverge.com/2018/3/20/17138854/cambridge-analytica-facebook-data-trump-campaign-psychographic-microtargeting

Chrisafis, Angelique: "France hit by day of protest as security forces fire teargas at taxi strike." *The Guardian,* 26 Jan 2016. Available at: https://www.theguardian.com/world/2016/jan/26/french-taxi-drivers-block-paris-roads-in-uber-protest

Clark, Andrew E.; Flèche, Sarah; Layard, Richard; Powdthavee, Nattavudh & Ward, George: *The Origins of Happiness: The Science of Well-Being over the Life Course* (Princeton University Press: 2018).

Clark, Gregory: A F*arewell to Alms – A Brief Economic History of the World* (Princeton University Press: 2007).

Clark, Gregory: Review Essay: The Enlightened Economy. An Economic History of Britain, 1700–1850 by Joel Mokyr. *Journal of Economic Literature,* March 2012, 85–95. Available at: http://faculty.econ.ucdavis.edu/faculty/gclark/Book_Reviews/Mokyr%20JEL%202012.pdf

Clayton, R. B.: The third wheel: The impact of Twitter use on relationship infidelity and divorce. *Cyberpsychology, Behavior, and Social Networking,* July 2014, 425–430. Available at: https://www.ncbi.nlm.nih.gov/pubmed/24690067

Clayton, R. B., Nagurney, A., & Smith, J. R.: "Cheating, breakup, and divorce: Is Facebook use to blame?" *Cyberpsychol-*

ogy, Behavior, and Social Networking, October 2013, 717–720. Available at: https://www.liebertpub.com/doi/abs/10.1089/cyber.2012.0424?journalCode=cyber

Clinton, Hillary: Campain Speech, Keene State University, 25 Jan 1996. Available at: https://www.c-span.org/video/?69606-1/mrs-clinton-campaign-speech

Collins, Randall: Keskiluokan työn loppu: Pakotietä ei ole. Wallerstein, Immanuel; Collins, Randall; Mann, Michael; Derluguian, Georgi & Calhoun, Craig: *Onko kapitalismilla tulevaisuutta?* (Gaudeamus/Vastapaino: 2014; orig. 2013).

Collishaw, Stephan: Annual Research Review: Secular trends in child and adolescent mental health. *The Journal of Child Psychology and Psychiatry* 12 Dec 2014. Available at: https://onlinelibrary-wiley-com.ezproxy.jyu.fi/doi/full/10.1111/jcpp.12372#jcpp12372-bib-0040

Cowen, Tyler: *The Great Stagnation: How America Ate All the Low-Hanging Fruit of Modern History, Got Sick, and Will (Eventually) Feel Better* (Dutton: 2011).

Cowen, Tyler: *Average is Over: Powering America Beyond the Age of Stagnation* (Penguin: 2013).

Cuddy, Alice: "The IQ of Europeans is dropping due to technology, say researchers," Euronews, 29 Dec 2017. Available at: http://www.euronews.com/2017/12/29/the-iq-of-europeans-is-dropping-due-to-technology-say-researchers

Dance, Gabriel J. X.; Confessore, Nicholas & LaForgia, Michael: "Facebook Gave Device Makers Deep Access to Data on Users and Friends," *The New York Times*, 3 Jun 2018. Available at: https://www.nytimes.com/interactive/2018/06/03/technology/facebook-device-partners-users-friends-data.html

Darvas, Zsolt: "This is the state of inequality in Europe," World Economic Forum, 4 May 2018. Available at: https://www.weforum.org/agenda/2018/05/european-income-inequality-begins-to-fall-once-again

Delaney, Kevin J.: "The Robot that takes your job should pay taxes, says Bill Gates," Quartz, 17 Feb 2017. Available at: https://qz.com/911968/bill-gates-the-robot-that-takes-your-job-should-pay-taxes/

Diamond, Jared: *Tykit taudit ja teräs: Ihmisen yhteiskuntien kohtalot.* Original English *Guns, Germs, and Steel: The Fates of Human Societies*, 1997. Translated into Finnish by Kimmo Pietiläinen (Terra Cognita: 2004).

Dilts, Elizabeth: "Apple says it looks out for kids, as investors cite phone 'addiction,'" Reuters, 8 Jan 2018. Available at: https://www.reuters.com/article/us-apple-shareholders-children/apple-says-it-looks-out-for-kids-as-investors-cite-phone-addiction-idUSKBN1EW0WS

Douglas, Emily: "Airbnb's former Global Head of Employee Experience on making values count," HRM, 28 Feb 2018. Available at: https://www.hrmonline.ca/hr-general-news/airbnbs-former-global-head-of-employee-experience-on-making-values-count-238375.aspx

Dowd, Maureen: "Elon Musk's Billion-Dollar Crusade to Stop the AI Apocalypse," *Vanity Fair*, April 2017.

Duhigg, Charles: "How Companies Learn Your Secrets," *The New York Times*, 16 Feb 2012. Available at: https://www.nytimes.com/2012/02/19/magazine/shopping-habits.html?_r=4&hp=&pagewanted=print

Dwoskin, Elizabeth; Harwell, Drew & Timberg, Craig: "Facebook had a closer relationship than it disclosed with the academic it called a liar," *The Washington Post*, 22 Mar 2018. Available at: https://www.washingtonpost.com/business/economy/facebook-had-a-closer-relationship-than-it-disclosed-with-the-academic-it-called-a-liar/2018/03/22/ca0570cc-2df9-11e8-8688-e053ba58f1e4_story.html?utm_term=.915a6247e387

Dwoskin, Elizabeth & Romm, Tony: "Facebook's rules for accessing user data lured more than just Cambridge Analytica," *The Washington Post*, 19 Mar 2018. Available at: https://www.washingtonpost.com/business/economy/face-books-rules-for-accessing-user-data-lured-more-than-just-cambridge-analytica/2018/03/19/31f6979c-658e-43d6-a71f-afdd8bf1308b_story.html?utm_term=.9af7014ced43

Dyson, George: *Turing's Cathedral: The Origins of the Digital Universe* (Penguin Books: 2013, orig. 2012).

Earth First! - Webpage, 6 Dec 2017. Available at: https://earthfirstjournal.org/newswire/2017/12/06/france-arson-at-la-casemate-fablab-center-for-scientific-technical-and-industrial-culture-of-grenoble/

Easterlin, Richard A.: Does Economic Growth Improve the Human Lot? Some Empirical Evidence. David, R. & Reder, R. (ed.) *Nations and Households in Economic Growth: Essays in Honor of Moses Abramovitz* (Academic Press: 1974). Available at: http://huwdixon.org/teaching/cei/Easterlin1974.pdf

Easterlin, Richard A.: Income and Happiness: Towards a Unified Theory, *The Economic Journal*, July 2001, 465–484. Available at: http://www.uvm.edu/pdodds/research/papers/others/2001/easterlin2001a.pdf

Easterlin, Richard A.; Angelescu McVey, Laura; Switek, Malgorzata; Sawangfa, Onnicha & Smith Zweig, Jacqueline: The happiness-income paradox revisited. *Proceedings of the National Academy of Sciences of the United States of America,* 28 Dec 2010, 22463–22468. Available at: https://www.ncbi.nlm.nih.gov/pmc/articles/PMC3012515/

The Economist: "Solving the Paradox," 21.9. 2000. Available at: http://www.economist.com/node/375522

The Economist, 12 Jan 2013.

The Economist: "The onrushing wave," 18 Jan 2014. *Available at:* https://www.economist.com/briefing/2014/01/18/the-onrushing-wave

The Economist: "The End of Moore's Law," 19 Apr 2015. Available at: https://www.economist.com/the-economist-explains/2015/04/19/the-end-of-moores-law

The Economist: "Automation and Anxiety," 25 Jun 2016a.

Available at: http://www.economist.com/news/special-report/21700758-will-smarter-machines-cause-mass-unemployment-automation-and-anxiety

The Economist: "From not working to neural networking," 25 Jun 2016b. Available at: https://www.economist.com/special-report/2016/06/25/from-not-working-to-neural-networking

The Economist: "The return of the machinery question," 25 Jun 2016c. Available at: http://www.economist.com/news/special-report/21700761-after-many-false-starts-artificial-intelligence-has-taken-will-it-cause-mass

The Economist: "China's tech trailblazers," 6 Aug 2016d. Available at: https://www.economist.com/leaders/2016/08/06/chinas-tech-trailblazers

The Economist: "Dawn of a new epoch?," 3 Sep 2016e. Available at: *http://www.economist.com/news/science-and-technology/21706227-people-may-have-propelled-earth-novel-episode-geological-time-dawn*

The Economist: "Test of Character," 29 Sep 2016f. https://www.economist.com/finance-and-economics/2016/09/29/tests-of-character

The Economist: "False news items are not the only problem besetting Facebook," 24 Nov 2016g. Available at: https://www.economist.com/news/business/21710835-falling-share-price-privacy-policy-and-advertising-metrics-also-keep-mark-zuckerberg

The Economist: "Technology is transforming what happens when a child goes to school," 22 Jul 2017a. Available at: https://www.economist.com/news/briefing/21725285-reformers-are-using-new-software-personalise-learning-technology-transforming-what-happens

The Economist: "Advances in AI are used to spot signs of sexuality," 9 Sep 2017b. Available at: https://www.economist.com/news/science-and-technology/21728614-machines-read-faces-are-coming-advances-ai-are-used-spot-signs

The Economist: "Researchers produce images of people's faces from their genomes," 9 Sep 2017c. Available at: https://www.economist.com/news/science-and-technology/21728613-facial-technology-makes-another-advance-researchers-produce-images-peoples

The Economist: "Are digital distractions harming labour productivity?," 7 Dec 2017d. Available at: https://www.economist.com/news/finance-and-economics/21732141-evidence-mixed-it-seems-clear-however-they-are-making-us

The Economist: "Teenagers are better behaved and less hedonistic nowadays," 10.1. 2018a. Available at: https://www.economist.com/international/2018/01/10/teenagers-are-better-behaved-and-less-hedonistic-nowadays

The Economist: "Cutting adolescents' use of social media will not solve their problems," 11 Jan 2018b. Available at: https://www.economist.com/leaders/2018/01/11/cutting-adolescents-use-of-social-media-will-not-solve-their-problems

The Economist: "Are programs better than people at predicting reoffending?," 17 Jan 2018c. Available at: https://www.economist.com/news/science-and-technology/21734986-short-answer-two-are-about-same-are-programs-better

The Economist: "How to tame the tech titans," 18 Jan 2018d. Available at: https://www.economist.com/leaders/2018/01/18/how-to-tame-the-tech-titans

The Economist: "Autonomous weapons are a game-changer," 25 Jan 2018e. Available at: https://www.economist.com/special-report/2018/01/25/autonomous-weapons-are-a-game-changer

The Economist: "WhatsApp suggests a cure for virality," 26 Jul 2018f. Available at: https://www.economist.com/leaders/2018/07/26/whatsapp-suggests-a-cure-for-virality

The Economist: "Against the tyranny of the majority," 4 Aug 2018g. Available at: https://www.economist.com/schools-brief/2018/08/04/against-the-tyranny-of-the-majority

Electronic Privacy Information Center: "Algorithms in the Criminal Justice System" (No date on webpage, noted as 25 Aug 2018). Available at: https://epic.org/algorithmic-transparency/crim-justice/

Eliot, T. S.: *The Rock* (orig. 1934). *The Poems of T.S. Eliot: Volume I* (Bloomsbury: 2015), 153–157.

Ellen MacArthur Foundation: "New Plastics Economy report offers blueprint to design a circular future for plastics," 19 Jan 2016. Available at: https://www.ellenmacarthurfoundation.org/news/new-plastics-economy-report-offers-blueprint-to-design-a-circular-future-for-plastics

Enbuske, Tuomas: "Kahdeksankymmentäluku oli helvetti," *Helsingin Sanomat* (*Nyt*), 10 Jan 2014. Available at: https://www.hs.fi/nyt/art-2000002701242.html

Epstein, Robert & Robertson, Ronald E.: The search engine manipulation effect (SEME) and its possible impact on the outcomes of elections. *Proceedings of the National Academy of Sciences of the United States of America.* 18 Aug 2015, E4512–E4521. Available at: http://www.pnas.org/content/112/33/E4512

EU Commission: "Antitrust: Commission fines Google €2.42 billion for abusing dominance as search engine by giving illegal advantage to own comparison shopping service," bulletin, 27 Jun 2017. Available at: http://europa.eu/rapid/press-release_IP-17-1784_en.htm

Evans, James A.: Electronic Publication and the Narrowing of Science and Scholarship. *Science,* July 2008, 395–399. Available at: http://science.sciencemag.org/content/321/5887/395

Facebook: "Bernie Sanders: 2016 Presidential Campaign." Available at: https://www.facebook.com/business/success/bernie-sanders

FBI: *Crime in the United States in 2013* (2014). Available at: https://ucr.fbi.gov/crime-in-the-u.s/2013/crime-in-the-u.s.-2013/offenses-known-to-law-enforcement/expanded-homicide

Federal Reserve Bank of St. Louis: *Civilian Population-Employ-*

ment Ratio (2018). Available at: https://fred.stlouisfed.org/series/EMRATIO

Fessler, Leah: "Researchers have replicated a notorious social experiment that claimed to explain the rise of fascism," Quartz, 15 Mar 2017. Available at: https://qz.com/932110/researchers-have-replicated-a-notorious-social-experiment-that-claimed-to-explain-the-rise-of-fascism/

Findicator: *Työllisyysaste* (2018). Available at: https://findikaattori.fi/fi/41

Fingas, Jon: "Twitter sold data access to researcher in Cambridge Analytica scandal," Endgadget, 29 Apr 2018. Available at: https://www.engadget.com/2018/04/29/twitter-sold-data-access-to-cambridge-analytica-scandal-researcher/

Finnish Institute for Educational Research: *Pisa-tutkimus: Suomalaisnuoret tyytyväisiä elämäänsä* (presentation slides 19 Apr 2017). Available at: https://ktl.jyu.fi/julkaisut/julkaisuluettelo/julkaisut/2017/KTL-D118

Finnish Institute for Health and Welfare: *Kouluterveyskyselyiden tulokset 2006-2017 (2017).* Available at: https://thl.fi/fi/web/lapset-nuoret-ja-perheet/tutkimustuloksia/kaikki-kouluterveyskyselyn-tulokset

The Finnish Medical Association: *Lukiolaisten, väestön ja lääkärien näkemyksiä terveydenhuollon tulevaisuudesta,* 10 Jul 2017. Available at: https://www.laakariliitto.fi/site/assets/files/1270/terveydenhuollon_digitalisaatio_tulokset_lukiolaiset.pdf

Fitzgerald, Rory; ESS Core Scientific Team; Ervasti, Heikki: *European Social Survey 2016: Suomen aineisto.*

Flynn, James R. & Shayer, Michael: IQ decline and Piaget: Does the rot start at the top? *Intelligence,* Jan-Feb 2018, 112–121. Available at: https://www.sciencedirect.com/science/article/pii/S0160289617302787

Ford, Martin: *Rise of the Robots: Technology and the Threat of a Jobless Future* (Basic Books: 2015).

Freedom House: "Freedom in the World 2018." Available at:

https://freedomhouse.org/report/freedom-world/freedom-world-2018

Freud, Sigmund: *Civilization and Its Discontents*. Original German *Das Unbehagen in der Kultur, 1930*. Translated into English by David McLintock (Penguin Books: 2002; orig. 1930).

Frey, Carl Benedikt & Osborne, Michael A.: *The Future of Employment: How Susceptible are Jobs to Computerisation?* 17 Sep 2013. Oxford Martin School. Available at: http://www.oxfordmartin.ox.ac.uk/downloads/academic/The_Future_of_Employment.pdf

Fukuyama, Francis: *Historian loppu ja viimeinen ihminen*. Original English *The End of History and The Last Man*, 1992. Translated into Finnish by Heikki Eskelinen (WSOY: 1992).

Fukuyama, Francis: *Our Posthuman Future: Consequences of the Biotechnology Revolution* (Farrar, Straus and Giroux: 2002).

Garcia-Martinez, Antonio: "How Trump Conquered Facebook – Without Russian Ads," *Wired*, 23 Feb 2018. Available at: https://www.wired.com/story/how-trump-conquered-facebookwithout-russian-ads/

Goethe, Johann Wolfgang von: *Faust: Murhenäytelmän ensimmäinen osa*. Original German *Faust. / Eine Tragödie 1808*. Translated into Finnish by Valter Juva (Otava: 1981a).

Goethe, Johann Wolfgang von: *Faust: Murhenäytelmän toinen osa*. Original German *Faust. / Der Tragödie zweiter Teil*, 1832. Translated into Finnish by Otto Manninen (Otava: 1981b).

Goldstein, Rebecca: *Incompleteness: The Proof and Paradox of Kurt Gödel* (W. W. Norton & Co: 2005).

Gorman, James: "A Sense of Where You Are," *The New York Times*, 29 Apr 2013. Available at: https://www.nytimes.com/2013/04/30/science/may-britt-and-edvard-moser-explore-the-brains-gps.html

Gorner, Jeremy: "Chicago police use 'heat list' as strategy to prevent violence," *Chicago Tribune*, 21 Aug 2013. Available at: http://articles.chicagotribune.com/2013-08-21/news/ctmet-heat-list-20130821_1_chicago-police-commander-andrew-papachristos-heat-list

Gourevitch, Victor (ed.): *Rousseau: The Discourses and other early political writings* (Cambridge University Press: 1997).

Grassegger, Hannes & Krogerus, Mikael: "The Data That Turned the World Upside Down," Vice Motherboard, 28 Jan 2017. Available at: https://motherboard.vice.com/en_us/article/mg9vvn/how-our-likes-helped-trump-win

Green, Joshua & Issenberg, Sasha: "Inside the Trump Bunker, with Days to Go," Bloomberg, 27 Oct 2016. Available at: https://www.bloomberg.com/news/articles/2016-10-27/inside-the-trump-bunker-with-12-days-to-go

Greenfield, Adam: "China's Dystopian Tech Could Be Contagious," *The Atlantic* 14 Feb 2018. Available at: https://www.theatlantic.com/technology/archive/2018/02/chinas-dangerous-dream-of-urban-control/553097/

Greenfield, Susan: *Mind Change: How digital technologies are leaving their mark on our brains* (Ebury Press: 2014).

Greenwald, Glenn; Grim, Ryan & Gallagher, Ryan: "Top-Secret Document Reveals NSA Spied on Porn Habits as Part of Plan to Discredit 'Radicalizers,'" Huffington Post, 26 Nov 2013. Available at: https://www.huffingtonpost.com/2013/11/26/nsa-porn-muslims_n_4346128.html

Gurven, Michael & Kaplan, Hillard: Longevity among Hunter-Gatherers: A Cross-Cultural Examination. *Population and Development Review*, June 2007, 321–365.

Haapaluoma-Höglund, Jaana: "Kone näkee ihmisen salaiset tunteet – toistaiseksi vain tutkijankammiossa," *Turun Sanomat* (STT), 15 Oct 2017. Available at: http://www.ts.fi/uutiset/kotimaa/3690898/Kone+nakee+ihmisen+salaiset+tunteet++toistaiseksi+vain+tutkijankammiossa

Hackett, Robert: "What to know about Ashley Madison hack," *Fortune*, 26 Aug 2015. Available at: http://fortune.com/2015/08/26/ashley-madison-hack/

Hadar, A. A.; Eliraz, D.; Lazarovits A.; Alyagon U; Zangen A.: Using longitudinal exposure to causally link smartphone usage to changes in behavior, cognition and right prefrontal neural activity. *Brain Stimulation*, Mar-Apr 2015. *Available at: https://www.brainstimjrnl.com/article/S1935-861X(15)00456-8/fulltext*

Haile, Tony: "What You Think You Know About the Web Is Wrong," *Time*, 9 Mar 2014. Available at: http://time.com/12933/what-you-think-you-know-about-the-web-is-wrong/

Haltiwanger, John: "Russia's military has new robot tank it says fights better than human soldiers," *Newsweek*, 9 Nov 2017. Available at: http://www.newsweek.com/russia-military-new-robot-tank-fights-better-human-soldiers-706836

Hämäläinen, Veli-Pekka: "Iso osa kouluista rajoittaa oppilaiden kännykänkäyttöä tavalla, joka hipoo sananvapauden rajoja," Yle Uutiset, 12.1. 2017. Available at: https://yle.fi/uutiset/3-9398509

Hansen, Lauren: "8 drivers who blindly followed their GPS into disaster," The Week, 7.5. 2013. Available at: http://theweek.com/articles/464674/8-drivers-who-blindly-followed-gps-into-disaster

Harding, Nick: "Eco anarchists: A new breed of terrorist?" *The Independent*, 18 May 2010. Available at: https://www.independent.co.uk/environment/eco-anarchists-a-new-breed-of-terrorist-1975559.html

Hartnett, Kevin: "To Build Truly Intelligent Machines, Teach Them Cause and Effect," Quanta Magazine, 15 May 2018. Available at: https://www.quantamagazine.org/to-build-truly-intelligent-machines-teach-them-cause-and-effect-20180515/

Harris, John: "In Sri Lanka, Facebook's Dominance has Cost Lives," *The Guardian*, 6 May 2018. Available at: https://www.theguardian.com/commentisfree/2018/may/06/sri-lanka-facebook-lives-tech-giant-poor-countries

Harris, Mark: "Inside the First Church of Artificial Intelligence," *Wired*, 15 Nov 2017. Available at: https://www.wired.com/story/anthony-levandowski-artificial-intelligence-religion/

Headrick, Daniel R.: *Power over Peoples: Technology, Environments, and Western Imperialism, 1400 to the Present* (Princeton University Press: 2010).

Henkel, Linda A.: Point-and-Shoot Memories: The Influence of Taking Photos on Memory for a Museum Tour. *Psychological Science*, February 2014, 396–402. Available at: http://journals.sagepub.com/doi/abs/10.1177/0956797613504438

Hellier, David: "From Rio to Paris – Uber is fighting battles across the globe," *The Guardian* 2 Oct 2015. Available at: https://www.theguardian.com/technology/2015/oct/02/uber-global-battles-from-rio-paris-amsterdam

Hern, Alex: "Far more than 87m Facebook users had data compromised, MPs told," *The Guardian*, 17 Apr 2018. Available at: https://www.theguardian.com/uk-news/2018/apr/17/facebook-users-data-compromised-far-more-than-87m-mps-told-cambridge-analytica

Hirvonen, Iida Sofia: "Tutkija Katri Saarikiveä harmittaa mielikuva, että Helsinki Challengessa vain pitchattaisiin," *Ylioppilaslehti* 11 Dec 2015. Available at: http://ylioppilaslehti.fi/2015/12/tutkija-katri-saarikivea-harmittaa-mielikuva-etta-helsinki-challengessa-vain-pitchattaisiin/

Hirvonen, Saara: "Ranska kieltää älypuhelimet kouluissa – täyskielto sai lainsäätäjiltä hyväksynnän," Yle Uutiset 31 Jul 2018. Available at: https://yle.fi/uutiset/3-10329813

Hobbes, Thomas: *Leviathan* (Cambridge University Press: 1996; orig. 1651).

Hobsbawm, Eric: The Machine Breakers. *Past & Present*, February 1952, 57–70. Available at: https://libcom.org/history/machine-breakers-eric-hobsbawm

Hobsbawm, Eric: *Äärimmäisyyksien aika: Lyhyt 1900-luku (1914–*

1991). Original English *The Age of Extremes: The Short Twentieth Century, 1914-1991*, 1994, Translated into Finnish by Pasi Junila (Vastapaino: 1999).

Honka, Niina: "Koulutuspolitiikan professori: Paras tapa saada Pisa-tulokset nousuun on hillitä nuorten älypuhelinten käyttöä," Yle Uutiset, 15 Jun 2018. Available at: https://yle.fi/uutiset/3-10251667?utm_source=facebook-share&utm_medium=social

Howard, Philip N.; Bolsover, Gillian; Kollanyi, Bence; Bradshaw, Samantha & Neudert, Lisa-Maria: *Junk News and Bots during the US Election: What Were Michigan Voters Sharing Over Twitter?* (Oxford University: 2017). Available at: http://comprop.oii.ox.ac.uk/wp-content/uploads/sites/89/2017/03/What-Were-Michigan-Voters-Sharing-Over-Twitter-v2.pdf

Human Rights Watch: *Making the Case: The Dangers of Killer Robots and the need for a Preemptive Ban* (2016). Available at: https://www.hrw.org/report/2016/12/09/making-case/dangers-killer-robots-and-need-preemptive-ban

Huxley, Aldous: *Uljas uusi maailma.* Original English *Brave New World*, 1932. Translated into Finnish by I. H. Orras (Tammi: 1978).

Hvistendahl, Mara: "Inside China's Vast New Experiment in Social Ranking," *Wired*, 14 Dec 2017. Available at: https://www.wired.com/story/age-of-social-credit/ILO & OECD: *The Labour Share in G20 Economies* (2015). Available at: https://www.oecd.org/g20/topics/employment-and-social-policy/The-Labour-Share-in-G20-Economies.pdf

Irwin, Neil: "Why Is Productivity So Weak? Three Theories." *The New York Times*, 28 Apr 2016. Available at: https://www.nytimes.com/2016/04/29/upshot/why-is-productivity-so-weak-three-theories.html

Isaac, Mike & Wakabayashi, Daisuke: "Russian Influence Reached 126 Million Through Facebook Alone," *The New York Times*, 30 Oct 2017. Available at: https://www.nytimes.

com/2017/10/30/technology/facebook-google-russia.html?action=click&contentCollection=Health&module=RelatedCoverage&region=Marginalia&pgtype=article

Jacobsen W.C. & Forste, R.: The wired generation: Academic and social outcomes of electronic media use among university students. *Cyberpsychology, Behavior, and Social Networking*, May 2011, 275–280. Available at: http://citeseerx.ist.psu.edu/viewdoc/download?doi=10.1.1.471.7633&rep=rep1&type=pdf

Jha, Alok: "Should we be taking a closer look at the potential dangers of nanotechnology?" *The Guardian*, 25 May 2011. Available at: https://www.theguardian.com/nanotechnology-world/dangers-of-nanotechnology-toxic

Johnson, Sally C.: *Psychological Evaluation of Theodore Kaczynski* (1998). Available at: http://paulcooijmans.com/psychology/unabombreport.html

Junco, R & Cotten, S. R.: No A 4 U: The relationship between multitasking and academic performance. *Computers & Education*, September 2012, 505–514. Available at: https://www.sciencedirect.com/science/article/pii/S036013151100340X

Junkkari, Marko & Liimatainen, Karoliina: "Suomalaiset uskovat töidensä säilyvän tulevaisuudessa – Tutkija: '2010-luvulla työpaikkojen tuhoutuminen on ollut vähäisempää kuin 1990- ja 2000-luvuilla,'" *Helsingin Sanomat*, 5 Sep 2016. Available at: https://www.hs.fi/talous/art-2000002919320.html

Junkkari, Marko & Teittinen, Paavo: "HS selvitti, mikä on hallituksen leikkauspolitiikkaan vaikuttava kiistelty koneäly Kooma – luotetaanko siihen liikaa?" *Helsingin Sanomat*, 20 Nov 2016. Available at: https://www.hs.fi/sunnuntai/art-2000004875641.html

Kaczynski, Theodore (Name of original author: FC): Industrial Society and its Future, *The Washington Post*, 22.9.1995. Available at: https://www.washingtonpost.com/wp-srv/national/longterm/unabomber/manifesto.text.htm

Kaczynski, Theodore: *Technological Slavery* (Feral House: 2010).

Kaczynski, Theodore: *Anti-Tech Revolution: Why and How* (Fitch & Madison: 2016).

Kahneman, Daniel; Krueger, Alan B; Schkade, David & Stone, Arthur A.: "Would you be happier if you were richer? A Focusing Illusion." *Science*, July 2006, 1908–1910. Available at: https://www.researchgate.net/publication/6974433_Would_You_Be_Happier_If_You_Were_Richer_A_Focusing_Illusion

Kahneman, Daniel & Deaton, Angus: High income improves evaluation of life but not emotional well-being. *Proceedings of the National Academy of Sciences of the United States of America*, 21 Sep 2010, 16489–16493. Available at: https://www.ncbi.nlm.nih.gov/pmc/articles/PMC2944762/

Kang, Cecilia: "Tech Industry Pursues a Federal Privacy Law, on Its Own Terms," *The New York Times*, 26 Aug 2018. Available at: https://www.nytimes.com/2018/08/26/technology/tech-industry-federal-privacy-law.html?action=click&module=MoreInSection&pgtype=Article&region=Footer&contentCollection=Technology

Kaplan, Jerry: *Humans Need Not Apply: A Guide to Wealth and Work in the Age of Artificial Intelligence* (Yale University Press: 2015).

Karpicke, Jeffrey D. & Blunt, Janell R.: Retrieval Practice Produces More Learning than Elaborative Studying with Concept Mapping. *Science*, 11. February 2011, 772–775. Available at: http://science.sciencemag.org/content/331/6018/772.full

Karpinski, Aryn C.; Kirschner, Paul A.; Ozer, Ipek; Mellott, Jennifer A.; Ochwo, Pius: "An exploration of social networking site use, multitasking, and academic performance among United States and European university students." *Computers in Human Behavior*, May 2013, 1182–1192. Available at: https://www.sciencedirect.com/science/article/pii/S0747563212002798?via%3Dihub

Karvala, Kreeta: "Suomeen tulossa jättimäinen sosiaaliturvan uudistus – näin poliitikot ottaisivat yleistuen käyttöön," 26 Mar 2018. Available at: https://www.iltalehti.fi/politiikka/201803252200835911_pi.shtml

Keen, Andrew: *Digital Vertigo: How Today's Online Social Revolution Is Dividing, Diminishing, and Disorienting Us* (St. Martin Griffin: 2013).

Kellaway, Lucy: Radio-column (no name), BBC World Service, 15 Aug 2016. Available at: http://www.bbc.co.uk/programmes/p043y84s

Kempas, Karla: "Miksei Kelan terveystiedoilla voisi olla hintaa, pohtii Juha Sipilä – puolueiden johtajat kohtasivat tekoälypaneelissa," *Helsingin Sanomat*, 15 Feb 2018. Available at: https://www.hs.fi/politiikka/art-2000005566896.html

Kharpal, Arjun: "Bill Gates wants to tax robots, but the EU says, 'no way, no way,'" CNBC News, 2 Jun 2017. Available at: https://www.cnbc.com/2017/06/02/bill-gates-robot-tax-eu.html

Kirn, Walter: "Little Brother is Watching," *The New York Times*, 15 Oct 2010. Available at: https://www.nytimes.com/2010/10/17/magazine/17FOB-WWLN-t.html?

Kivinen, Osmo & Hedman, Juha: Moniselitteiset Pisa-tulokset ja niiden ongelmalliset koulutuspoliittiset tulkinnat. *Politiikka*, 4/2017, 250–263.

Knight, Will: "'Info-mania' dents IQ more than marijuana," *New Scientist*, 22 Apr 2005. Available at: https://www.newscientist.com/article/dn7298-info-mania-dents-iq-more-than-marijuana/

Konrath, Sara, H; O'Brien, Edward H. & Hsing, Courtney: "Changes in Dispositional Empathy in American College Students Over Time: A Meta-Analysis." *Personality and Social Psychology Review*, May 2011, 180–198. Available at: http://journals.sagepub.com/doi/abs/10.1177/1088868310377395

Kontula, Osmo: *Finsex-tutkimus* (Väestöliitto: 2015). Available at: http://www.vaestoliitto.fi/tieto_ja_tutkimus/vaestontutkimuslaitos/seksologinen_tutkimus/suomalaisten-seksuaalisuus-finse/

Kopf, Dan: "Compelling new evidence that robots are taking jobs and cutting wages," Quartz, 28 Mar 2017. Available at: https://qz.com/943073/compelling-new-evidence-that-robots-are-taking-jobs-and-cutting-wages/

Kosinski, Michal; Stillwell, David & Graepel, Thore: Private traits and attributes are predictable from digital records of human behavior. *Proceedings of the National Academy of Sciences of the United States of America*, 9 Apr 2013, 5802–5805. Available at: http://www.pnas.org/content/110/15/5802

Koski, Olli & Husso, Kai (ed.): *Tekoälyajan työ: Neljä näkökulmaa talouteen, työllisyyteen, osaamiseen ja etiikkaan* (Ministry of Economic Affairs and Employment of Finland: 2018). Available at: http://julkaisut.valtioneuvosto.fi/bitstream/handle/10024/160931/19_18_TEM_Tekoalyajan_tyo_WEB.pdf

Kramer, Adam D. I.; Guillory, Jamie E. & Hancock, Jeffrey T. Experimental evidence of massive-scale emotional contagion through social networks, *Proceedings of the National Academy of Sciences of the United States of America*, 17 Jun 2014, 8788–8790. Available at: http://www.pnas.org/content/pnas/111/24/8788.full.pdf

Kross, E.; Verduyn, P.; Demiralp, E.; Park, J.; Lee, D. S.; Lin, N.; Shablack, H.; Jonides, J. & Ybarra, O.: Facebook use predicts declines in subjective well-being in young adults. *PloS One*, 14 Aug 2013. Available at: https://www.ncbi.nlm.nih.gov/pubmed/23967061

Krugman, Paul: "How Did Economists Get It So Wrong?" *The New York Times*, 2 Sep 2009. Available at: https://www.nytimes.com/2009/09/06/magazine/06Economic-t.html

Kunttu, Kristina; Pesonen, Tommi & Saari, Juhani: *Korkeakouluopiskelijoiden terveystutkimus 2016* (YTHS), 30–38. Available at: http://www.yths.fi/filebank/4237KOTT_2016_korjattu_final_0217.pdf

Kurzweil, Ray: The *Singularity Is Near: When Humans Transcend Biology* (Viking: 2005).

Kwok, Sharon: "Alibaba tops e-commerce market share while facing fresh competition in China," Marketing, 11 Jul 2018. Available at: http://www.marketing-interactive.com/alibaba-tops-e-commerce-market-share-while-facing-fresh-competition-in-china/

La Casemate center website 21 Nov 2017. Available at: https://www.ecsite.eu/members/members-share/news/fire-la-casemate

Laine, Linda: "Kiinassa poliisit tunnistavat rikollisia tarkkailulasien avulla – poliittisten toisinajattelijoiden ja vähemmistöjen pelätään joutuvan tehokkaamman tarkkailun kohteiksi," *Helsingin Sanomat* 8 Feb 2018. Available at: https://www.hs.fi/ulkomaat/art-2000005558570.html

Lamothe, Daniel: "Pentagon agency wants drones to hunt in packs, like wolves," *The Washington Post,* 23 Jan 2015. https://www.washingtonpost.com/news/checkpoint/wp/2015/01/23/pentagon-agency-wants-drones-to-hunt-in-packs-like-wolves/?utm_term=.dd8b3f2b3a0e

Larson, Jeff; Mattu, Suria; Kirchner, Lauren; Angwin, Julia: "How We Analyzed the COMPAS Recidivism Algorithm," ProPublica, 23 May 2016. Available at: https://www.propublica.org/article/how-we-analyzed-the-compas-recidivism-algorithm

Lawler, Peter Augustine: "Thielism: Trumpian Elitism." *National Review,* 12 Jan 2017. Available at: https://www.nationalreview.com/corner/peter-thiel-libertarian-technology-donald-trump-supporter-elitist/

Layard, Richard: *Happiness: Lessons from a New Science* (Penguin Press: 2005).

Lazareva, Inna: "Bored of Facebook news? You're not alone: study," Reuters, 14 Jun 2018. Available at: https://www.reuters.com/article/us-media-news-survey/bored-of-facebook-news-youre-not-alone-study-idUSKBN1J939V

Leavitt, David: *The Man Who Knew Too Much: Alan Turing and the Invention of the Computer* (W. W. Norton & Company: 2006).

Leiva, L.; Böhmer, M; Gehring, S & Krüger, A.: "Back to app: the costs of mobile application interruptions." *Proceedings of the Fourteenth International Conference of Human-Computer Interaction with Mobile Devices and Services – mobile HCI,* 2012, 291–294. Available at: https://www.semanticscholar.org/paper/Back-to-the-app%3A-the-costs-of-mobile-application-i-Leiva-B%C3%B6hmer/270ab1ce84fd8ef46f5e0099a78742e19b8e0573

Lepp, A.; Barkley, J. E. & Karpinski, A. C.: The relationship between cell phone use, academic performance, anxiety, and satisfaction with life in college students. *Computers in Human Behavior, 2014 helmikuu,* 343–350. Available at: https://www.sciencedirect.com/science/article/pii/S0747563213003993?via%3Dihub

Levitsky, Steven & Ziblatt, Daniel: *How Democracies Die: What History Reveals about our Future* (Viking: 2018).

Levy, David: *Love and Sex with Robots: The Evolution of Human–Robot Relationships* (HarperCollins Publishers: 2009; orig. 2007).

Levin, Sam; Wong, Julia Carrie & Harding, Luke: "Facebook backs down from 'napalm girl' censorship and reinstates photo," *The Guardian,* 9.9. 2016. Available at: https://www.theguardian.com/technology/2016/sep/09/facebook-reinstates-napalm-girl-photo

Levin, Sam: "Facebook admits it poses mental health risk – but says using site more can help," *The Guardian* 15 Dec 2017. Available at: https://www.theguardian.com/technology/2017/dec/15/facebook-mental-health-psychology-social-media

Lewis, Paul: "California's would-be governor prepares for battle against job-killing robots," *The Guardian,* 5 Jun 2017 (a).

Available at: https://www.theguardian.com/us-news/2017/jun/05/gavin-newsom-governor-election-silicon-valley-robots

Lewis, Paul: "'Our Minds Can Be Hijacked': the tech insiders who fear a smartphone dystopia," *The Guardian*, 6 Oct 2017 (b). Available at: https://www.theguardian.com/technology/2017/oct/05/smartphone-addiction-silicon-valley-dystopia

Lewis, Paul: "'Utterly horrifying': ex-Facebook insider says covert data harvesting was routine," *The Guardian*, 20 Mar 2018. Available at: https://www.theguardian.com/news/2018/mar/20/facebook-data-cambridge-analytica-sandy-parakilas

Liimatainen, Karoliina: "Pörssiyhtiö Tieto otti johtoryhmänsä jäseneksi tekoälyn – HS kysyi Tiedolta, onko Alicia T vain mainostemppu," *Helsingin Sanomat*, 19 Oct 2016. Available at: https://www.hs.fi/talous/art-2000002926147.html

Linkola, Pentti: *Johdatus 1990-luvun ajatteluun* (WSOY: 1989).

Loh, Kep Kee & Kanai, Ryota: How Has the Internet Reshaped Human Cognition? *The Neuroscientist*, October 2016, 506–520. Available at: https://www.ncbi.nlm.nih.gov/pubmed/26170005

Loh, Kep Kee & Kanai, Ryota: Higher Media Multi-Tasking Activity Is Associated with Smaller Gray-Matter Density in the Anterior Cingulate Cortex. *PLOS One*, September 2014. Available at: http://journals.plos.org/plosone/article?id=10.1371/journal.pone.0106698

Lohr, Steve: "A.I. Is Doing Legal Work. But It Won't Replace Lawyers, Yet," *The New York Times*, 19 Mar 2017. Available at: https://www.nytimes.com/2017/03/19/technology/lawyers-artificial-intelligence.html

Luotonen, Anniina & Lehtovaara, Mimma: "EU ryhtyy metsästämään veroeuroja amerikkalaisilta digijäteiltä – 'Tämä ei ole USA-vastainen vero,'" *Keskisuomalainen* (STT), 21 Mar 2018. Available at: https://www.ksml.fi/talous/EU-ryhtyymets%C3%A4st%C3%A4m%C3%A4%C3%A4n-veroeuroja-amerikkalaisilta-digij%C3%A4teilt%C3%A4-%E2%80%93-T%C3%A4m%C3%A4-ei-ole-USA-vastainen-vero/1126117

Lynge, E.: Unemployment and cancer: a literature review. *IARC Scientific Publications*, January 1997, 343–351. Available at: https://www.ncbi.nlm.nih.gov/pubmed/9353675

Madrigal, Alexis: "Drone Swarms Are Going to Be Terrifying and Hard to Stop," *The Atlantic*, 7 Mar 2018a. Available at: https://www.theatlantic.com/technology/archive/2018/03/drone-swarms-are-going-to-be-terrifying/555005/

Madrigal, Alexis: "What Took Facebook So Long?" *The Atlantic*, 18 Mar 2018b. Available at: https://www.theatlantic.com/technology/archive/2018/03/facebook-cambridge-analytica/555866/

Mark, Gloria; Gudith, Daniela & Klocke, Ulrich: The Cost of Interrupted Work: More Speed and Stress. *Proceedings of the 2008 Conference on Human Factors in Computing Systems*, Florence, Italy, 10–15 Apr 2008. Available at: https://www.ics.uci.edu/~gmark/chi08-mark.pdf

Markoff, John: "Computer Wins on 'Jeopardy!': Trivial, It's Not," *The New York Times*, 16 Feb 2011. Available at: https://www.nytimes.com/2011/02/17/science/17jeopardy-watson.html

Mascarenhas, Hyacinth: "Stephen Hawking: AI could 'develop a will of its own' in conflict with ours that 'could destroy us,'" *International Business Times*, 8 Nov 2017. Available at: https://www.ibtimes.co.uk/stephen-hawking-ai-could-develop-will-its-own-conflict-ours-that-could-destroy-us-1646352

Matikainen, Jenny: "Kuka viittaa, kuka nuokkuu? Kiinalaislukiossa kasvoja tunnistavat kamerat valvovat oppilaiden jokaista ilmettä ja asentoa," Yle Uutiset, 7 Aug 2018. Available at: https://yle.fi/uutiset/3-10296993

Matz, S. C.; Kosinski, M.; Nave, G.; Stillwell, D. J.: Psychological targeting as an effective approach to digital mass persuasion. *Proceedings of the National Academy of Sciences of the United States of America*, 13 Nov 2017. Available at: http://www.pnas.org/content/early/2017/11/07/1710966114

McCormack, John: "The Election Came Down to 77,744 Votes in Pennsylvania, Wisconsin, and Michigan (Updated)" *The Weekly Standard*, 10 Nov 2016. Available at: https://www.weeklystandard.com/john-mccormack/the-election-came-down-to-77-744-votes-in-pennsylvania-wisconsin-and-michigan-updated

McCormick, Emily: "Instagram Is Estimated to Be Worth More than $100 Billion," Bloomberg, 25 Jun 2018. Available at: https://www.bloomberg.com/news/articles/2018-06-25/value-of-facebook-s-instagram-estimated-to-top-100-billion

McDermott, Drew: *Artificial Intelligence and Consciousness* (Yale University: 2007). Available at: http://www.cs.yale.edu/homes/dvm/papers/conscioushb.pdf

McFarland, Matt: "Elon Musk: 'With artificial intelligence we are summoning the demon,'" *The Washington Post*, 24 Oct 2014. Available at: https://www.washingtonpost.com/news/innovations/wp/2014/10/24/elon-musk-with-artificial-intelligence-we-are-summoning-the-demon/?utm_term=.57c4eff0e4bf

McKew, Molly: "How Liberals Amped Up a Parkland Shooting Conspiracy Theory," *Wired*, 27 Feb 2018. Available at: https://www.wired.com/story/how-liberals-amped-up-a-parkland-shooting-conspiracy-theory/

McKinsey Global Institute: *The four global forces breaking all the trends* (2015). Available at: https://www.mckinsey.com/business-functions/strategy-and-corporate-finance/our-insights/the-four-global-forces-breaking-all-the-trends

McKinsey Global Institute: *Harnessing automation for a future that works* (2017a). Available at: https://www.mckinsey.com/featured-insights/digital-disruption/harnessing-automation-for-a-future-that-works

McKinsey Global Institute: *Shaping the future of work in Europe's 9 digital front-runner countries* (2017b). Available at: https://www.mckinsey.com/featured-insights/europe/shaping-the-future-of-work-in-europes-nine-digital-front-runner-countries

McLaughlin, Timothy: "Facebook blocks accounts of Myanmar's top general, other military leaders," *Washington Post*, 27 Aug 2018 Available at: https://www.washingtonpost.com/world/asia_pacific/facebook-blocks-accounts-of-myanmars-top-general-other-military-leaders/2018/08/27/da1ff440-a9f6-11e8-9a7d-cd30504ff902_story.html?utm_term=.03e73527de4b

McLuhan, Marshall: *Understanding Media: The Extensions of Man* (Gingko Press 2003; orig. 1964).

Medeiros, João: "Stephen Hawking: 'I fear AI may replace humans altogether,'" *Wired*, 28 Nov 2017. Available at: https://www.wired.co.uk/article/stephen-hawking-interview-alien-life-climate-change-donald-trump

Menger, Karl: *Reminiscences of the Vienna Circle and the Mathematical Colloquium* (Springer: 1994).

Metz, Cade: "Finally, neural networks that actually work," *Wired*, 21 Apr 2015. Available at: https://www.wired.com/2015/04/jeff-dean/https://www.scientificamerican.com/article/why-the-human-brain-project-went-wrong-and-how-to-fix-it/

Metz, Cade: "Inside Libratus, the Poker AI that out-bluffed the best humans," *Wired*, 2 Jan 2017a. Available at: https://www.wired.com/2017/02/libratus/

Metz, Cade: "Google's dueling neural networks spar to get smarter, no humans required," *Wired*, 11 Apr 2017b. Available at: https://www.wired.com/2017/04/googles-dueling-neural-networks-spar-get-smarter-no-humans-required/

Metz, Cade: "Pentagon Wants Silicon Valley's Help on AI," *The New York Times*, 15 Mar 2018. Available at:https://www.nytimes.com/2018/03/15/technology/military-artificial-intelligence.html

Metz, Cade & Singer, Natasha: "Newspaper Shooting Shows Widening Use of Facial Recognition by Authorities," *The New York Times*, 29 Jun 2018. Available at: https://www.nytimes.com/2018/06/29/business/newspaper-shooting-facial-recognition.html

Metzger, Max: "9.2 million medical records for sale on darkweb," *SC Magazine*, 29 Jun 2016. Available at: https://www.scmagazineuk.com/92-million-medical-records-for-sale-on-darkweb/article/530456/

Michaels, Guy; Natraj, Ashwini & Van Reenen, John: Has ICT polarized skill demand? Evidence from 11 countries over 25 years. *Review of Economics and Statistics*, August 2014, 60–77. Available at: http://eprints.lse.ac.uk/46830/1/Michaels_Natraj_VanReenen_Has-ICT-polarized-skill-demand_2014.pdf

Miller, James: *Singularity Rising: Surviving and Thriving in a Smarter, Richer, and More Dangerous World*, (BenBella Books: 2012).

Mishina, Kaisa; Tiiri, Elina; Lempinen, Lotta; Sillanmäki, Lauri; Kronström, Kim & Sourander, Andre: Time trends of Finnish adolescents' mental health and use of alcohol and cigarettes from 1998 to 2014. *European Child & Adolescent Psychiatry*, April 2018, Available at: https://www.researchgate.net/publication/324817583_Time_trends_of_Finnish_adolescents%27_mental_health_and_use_of_alcohol_and_cigarettes_from_1998_to_2014

Mitrunen, Matti: *Työmarkkinoiden polarisaatio Suomessa* (VATT: 2013). Available at: https://vatt.fi/artikkeli/-/asset_publisher/tyomarkkinoiden-polarisaatio-voimistuu

Moisala, Mona: *Brain activations related to attention and working memory and their association with technology-mediated activities* (Dissertation, Helsinki University: 2017).

Mokyr, Joel: *The Lever of Riches – Technological Creativity and Eco-*

nomic Progress (Oxford University Press: 1990).

Mokyr, Joel: "Technological Inertia in Economic History." *The Journal of Economic History*, June 1992, 325–338.

Mokyr, Joel: *The Enlightened Economy: An Economic History of Britain, 1700-1850* (Yale University Press: 2009).

Moody, Oliver: "Dumb and dumber: why we're getting less intelligent," *The Times*, 12 Jun 2018. Available at: https://www.thetimes.co.uk/edition/news/dumb-and-dumber-why-were-getting-less-intelligent-80k3bl83v

Morris, Ian: *Why the West Rules – For Now. The Patterns of History, and What They Reveal About the Future* (Farrar, Straus and Giroux: 2010).

Murphy, Heather: "Genealogists Turn to Cousins' DNA and Family Trees to Crack Five More Cold Cases," *The New York Times*, 27 Jun 2018. Available at: https://www.nytimes.com/2018/06/27/science/dna-family-trees-cold-cases.html

Mustonen, Ville: "Voiko evoluutiota ennustaa," *Tieteessä tapahtuu*, 2/2018, 46–49. Available at: https://journal.fi/tt/article/view/69936/31035

NASA: "Scientific Consensus: Earth's Climate is Warming." Available at: https://climate.nasa.gov/scientific-consensus/

Newlon, Cara: "Study: Millennials find technology dehumanizing," *USA Today*, 23 Oct 2013. Available at: https://eu.usatoday.com/story/news/nation/2013/10/22/study-millennials-tech-dehumanizing/3151693/

Nietzsche, Friedrich: *Näin puhui Zarathustra: Kirja kaikille ja ei kenellekään*. Original German *Also sprach Zarathustra – Ein Buch für Alle und Keinen*, 1883–1891. Translated into Finnish by J. A. Hollo (Otava: 1995; orig. 1883–1891).

Nisbet, Robert: *History of the Idea of Progress* (Basic Books: 1980).

Nixon, Dan: "Is the Economy suffering from the crisis of attention?" Englannin pankin Bank Underground - blog, 24 Nov 2017. Available at: https://bankunderground.co.uk/2017/11/24/is-the-economy-suffering-from-the-crisis-of-attention/

Nuotio, Tanja: "Jopa joka 10. nuori on psykiatrisessa hoidossa," *Kainuun Sanomat*, 20 Feb 2018.

O'Brien, Chris: "Meet the presidential candidate who's using the internet to reinvent French politics," VentureBeat, 8 Jan 2017: Available at: https://venturebeat.com/2017/01/08/meet-the-french-presidential-candidate-whos-using-the-internet-to-reinvent-politics/

OECD: *Students, Computers, and Learning: Making the Connection* (2015). Available at: https://www.keepeek.com//Digital-Asset-Management/oecd/education/students-computers-and-learning_9789264239555-en#.WpXRCejFI2w

OECD: *Pisa 2015 Results, Volume III: Student's Well-Being* (2017a), 7. Figure III.7.1. Available at: https://read.oecd-ilibrary.org/education/pisa-2015-results-volume-iii/change-through-2003-2012-and-2015-in-students-sense-of-belonging-at-school_9789264273856-graph29-en#page1

OECD: *Pisa 2015 Results, Volume III: Student's Well-Being* (2017b). Available at: https://www.oecd-ilibrary.org/docserver/9789264273856-11-en.pdf expires=1526839568&id=id&accname=guest&checksum=A7E0FDBE6E0436A47DCB9AD56BF66029

OECD: *Level of GDP per Capita and Productivity 1970–2016* (2018a). Available at: https://stats.oecd.org/Index.aspx?DataSetCode=PDB_LV

OECD: *Average usual weekly hours worked on the main job* (2018b). Available at: https://stats.oecd.org/Index.aspx?DataSetCode=AVE_HRS

Ogburn, William F.: *Social Change with Respect to Culture and Original Nature* (B. W. Huebsch: 1922).

Olfson, Mark; Blanco, Carlos & Wang, Shuai: National trends in the Mental Health Care of Children, Adolescents, and Adults by Office-Based Physicians. *JAMA Psychiatry*, January 2014, 81–90. Available at: https://jamanetwork-com.ezproxy.jyu.fi/

journals/jamapsychiatry/fullarticle/1784344

O'Neil, Cathy: *Weapons of Math Destruction: How Big Data Increases Inequality and Threatens Democracy* (Crown Books: 2016).

Ophir, E.; Nass, C & Wagner, A. D.: Cognitive control in media multitaskers. *Proceedings of the National Academy of Sciences of the United States of America,* 15. September 2009, 15583–15587. Available at: https://www.ncbi.nlm.nih.gov/pmc/articles/PMC2747164/

Orwell, George: *Vuonna 1984.* Original English *Nineteen Eighty-Four,* 1949. Translated into Finnish by Raija Mattila (WSOY: 2014).

Pääkkönen, Hannu & Hanifi, Riikka: *Ajankäytön muutokset 2000-luvulla* (Tilastokeskus: 2011). Available at: http://www.tilastokeskus.fi/tup/julkaisut/tiedostot/isbn_978-952-244-331-1.pdfPajarinen, Mika & Rouvinen, Petri: *Computerization Threatens One Third of Finnish Employment,* 13 Jan 2014, ETLA Economic Research. Available at: https://www.etla.fi/en/publications/computerization-threatens-finnish-employment/

Palihapitiya, Chamath: esiintyminen Stanfordin yliopistossa 13 Nov 2017. Available at: https://www.youtube.com/watch?v=PMotykw0SIk&t=1630s

Pantsu, Pekka: "Onnellisuusprofessori: suomalaisten onnellisuus on 1970-luvun tasolla – YK: Suomi on maailman onnellisin maa," Yle Uutiset, 21 Mar 2018. Available at: https://yle.fi/uutiset/3-10126051

Parker, Alex M.: "The Great Debates," *The Atlantic,* October 2004. Available at: https://www.theatlantic.com/magazine/archive/2004/10/the-great-debates/303593/

Parviainen, Aapo & Konttinen, Matti: "Nobelisti Holmström myönteisistä talousnäkymistä: 'Pitää katsoa pidemmälle, mitä huolia horisontissa näkyy,'" Yle Uutiset, 16 Jun 2017. Available at: https://yle.fi/uutiset/3-9675907

Pasternack, Alex: "Deadly Violence erupts in Kenya following contested, fake news-fueled election," *Fast Company*, 12 Aug 2017. Available at: https://www.fastcompany.com/40452960/kenya-election-deadly-violence-fake-news

Pearl, Judea: *The Book of Why: The New Science of Cause and Effect* (Basic Books: 2018).

Penney, Jonathon W: Internet Surveillance, Regulation, and Chilling Effects Online: A Comparative Case Study. *Internet Policy Review*, 26 May 2017. Available at: https://policyreview.info/articles/analysis/internet-surveillance-regulation-and-chilling-effects-online-comparative-case

Perera, Ayeshea: "The people trying to fight fake news in India," BBC, 24 Jul 2017. Available at: http://www.bbc.com/news/world-asia-india-40657074

Perkiö, Johanna & Pulkka, Ville-Veikko: Perustulo ja työllisyyspolitiikka. Kajanoja, Jouko (ed.): *Työllisyyskysymys* (Into: 2018).

Piketty, Thomas: *Pääoma 2000-luvulla*. Original French *Le Capital au XXIe siècle*, 2013. Translated into Finnish by Marja Ollila ja Maarit Tillman-Leino (Into: 2016).

Pilke, Antti: "Kysely: Suomalaiset kaipaavat takaisin 1980-luvulle," Yle Uutiset, 29 Dec 2013. Available at: https://yle.fi/uutiset/3-7003811

Pilling, David: *The Growth Delusion: Wealth, Poverty, and the Well-Being of Nations* (Bloomsbury: 2018).

Pinker, Steven: *The Better Angels of Our Nature – Why Violence has Declined* (Penguin: 2012; orig. 2011).

Platon: *Phaedrus. Plato in Twelve Volumes* (Harvard University Press: 1925). Available at: http://www.perseus.tufts.edu/hopper/text?doc=Perseus%3Atext%3A1999.01.0174%3Atext%3DPhaedrus%3Apage%3D275

Polanyi, Karl: *Suuri Murros: Aikakautemme poliittiset ja taloudelliset juuret* (Vastapaino: 2008, orig. 1944).

Popper, Ben: "Google's project to 'cure death,' Calico, announces $1,5 billion research center," The Verge, 3 Sep 2014. Available at: https://www.theverge.com/2014/9/3/6102377/google-calico-cure-death-1-5-billion-research-abbvie

Poutanen, Pauli: "Vihapuhetta kitkevä laki kuohuttaa Saksassa – Twitter sensuroi jopa satiirilehden tilin," *Iltalehti* 7.1. 2018. Available at: http://www.iltalehti.fi/ulkomaat/201801072200651916_ul.shtml

Price, Emily: "Bill Gates' Plan to Tax Robots Could Become a Reality in San Francisco," *Fortune*, 5 Sep 2017. Available at: http://fortune.com/2017/09/05/san-francisco-robot-tax/

Prime Minister's Office: *Ratkaisujen Suomi: Pääministeri Juha Sipilän hallituksen strateginen ohjelma*. Finnish Government Publication 10/2015.

Pulkka, Ville-Pekka: "Suomalaiset eivät usko työn loppuun," Nation Foresight Cooperation website, 6 Oct 2017. Available at: http://foresight.fi/ville-veikko-pulkka-suomalaiset-eivat-usko-tyon-loppuun/

Qiu, Linda: "Fingerprints of Russian Disinformation: From AIDS to Fake News," *The New York Times*, 12 Dec 2017. Available at: https://www.nytimes.com/2017/12/12/us/politics/russian-disinformation-aids-fake-news.html?module=ArrowsNav&contentCollection=Politics&action=keypress&region=FixedLeft&pgtype=article

Raeste, Juha-Pekka: "Menikö maailma liian monimutkaiseksi? Tämän takia retropuhelin Nokia 3310:n suosio on 'murskaava,'" *Helsingin Sanomat*, 8 Mar 2017: Available at: https://www.hs.fi/paivanlehti/08032017/art-2000005117056.html

Raittila, Hannu: Ikuinen kaupunki. *Yhteiskuntapolitiikka* 4/2004, 429–432. Available at: https://julkari.fi/bitstream/handle/10024/100941/404raittila.pdf?sequence=1;

Rees, Martin: *Our Final Century: Will Our Civilization Survive the Twenty-first Century* (Arrow Books: 2004).

Reuters: "Young adults say technology can be dehumanizing –poll," 17 Oct 2013. Available at: http://www.reuters.

com/article/us-technology-poll/young-adults-say-technology-can-be-dehumanizing-poll-idUSBRE99G0J020131017

Reuters: "Selena Gomez reveals Instagram addiction, low self-esteem," 16 Mar 2017. Available at: https://www.reuters.com/article/us-people-selenagomez/selena-gomez-reveals-instagram-addiction-low-self-esteem-idUSKBN16N2NZ

RoBhat Labs: "An Analysis of Propaganda Bots on Twitter," Medium, 31 Oct 2017. Available at: https://medium.com/@robhat/an-analysis-of-propaganda-bots-on-twitter-7b7ec57256ae

Rosen, Larry D.; Carrier, Mark L. & Cheever, Nancy A.: Facebook and texting made me do it: Media-induced task-switching while studying. *Computers in Human Behavior, May 2013, 948–958.* Available at: https://www.sciencedirect.com/science/article/pii/S0747563212003305

Roser, Max: "Democracy," Our World in Data, 2018. Available at: https://ourworldindata.org/democracy

Rousseau, Jean-Jacques: *Discourse on Inequality: On the Origin and Basis of Inequality Among Men* (Floating Press 2009; orig. 1755).

Roy, Jessica: "The Rapture of Nerds," *Time,* 17 Apr 2014. Available at: http://time.com/66536/terasem-trascendence-religion-technology/

Rubin, Peter: "Coming Attractions: The Rise of VR Porn," *Wired,* 17 Apr 2018. Available at: https://www.wired.com/story/coming-attractions-the-rise-of-vr-porn/

Russell, Jon: "Google's AlphaGo AI wins three-match series against the world's best Go player," TechCrunch, 25 May 2017. Available at: https://techcrunch.com/2017/05/24/alphago-beats-planets-best-human-go-player-ke-jie/

Russell, Stuart: "Take a stand on AI weapons," *Nature,* 27 May 2015. Available at: https://www.nature.com/news/robotics-ethics-of-artificial-intelligence-1.17611

Saari, Miia: "Ystävät 24/7," *Suomen Kuvalehti,* 17 Dec 2017.

Available at: https://suomenkuvalehti.fi/jutut/kotimaa/snapchat-kannustaa-teineja-pitamaan-ystavyyssuhteitaylla-kellon-ympari/

Salvadori, Massimo (ed.): *Modern Socialism* (Macmillan & Co: 1968).

Sample, Ian: "Scientists plan huge European AI hub to compete with US," *The Guardian*, 23 Apr 2018. Available at: https://www.theguardian.com/science/2018/apr/23/scientists-plan-huge-european-ai-hub-to-compete-with-us

Sana, Faria; Weston, Tina & Cepeda, Nicholas J.: Laptop multitasking hinders classroom learning for both users and nearby peers. *Computers & Education*, March 2013, 24–31. Available at: https://www.sciencedirect.com/science/article/pii/S0360131512002254

Scharre, Paul: *Army of None: Autonomous Weapons and the Future of War* (W. W. Norton: 2018).

Schneider, Leonid: "The laborious delivery of Markram's brainchild," For Better Science - blog , 15 Jul 2016. Available at: https://forbetterscience.com/2016/07/15/the-laborious-delivery-of-markrams-brainchild/

Schneier, Bruce: "How the NSA Threatens National Security," Schneier on Security -blog, 6 Jan 2014. https://www.schneier.com/essays/archives/2014/01/how_the_nsa_threaten.html

Schneier, Bruce: "It's Not Just Facebook. Thousands of Companies are Spying on You," Schneier on Security -blog, 26 Mar 2018. Available at: https://www.schneier.com/essays/archives/2018/03/its_not_just_faceboo.html

Schonfeld, Erick: "Eric Schmidt Tells Charlie Rose Google Is 'Unlikely' To Buy Twitter and Wants To Turn Phones Into Tvs," TechCrunch, 7 Mar 2009. Available at: https://techcrunch.com/2009/03/07/eric-schmidt-tells-charlie-rose-google-is-unlikely-to-buy-twitter-and-wants-to-turn-phones-into-tvs/

Shakya, Holly B. & Christakis, Nicholas A.: Association of Facebook Use With Compromised Well-Being: A Longitudinal Study. *American Journal of Epidemiology*, February 2017, 203–211. Available at: https://academic.oup.com/aje/article/185/3/203/2915143

Shane, Scott; Metz, Cade & Wakabayashi, Daisuke: "How a Pentagon Contract Became an Identity Crisis for Google," *The New York Times*, 30 May 2018. Available at: https://www.nytimes.com/2018/05/30/technology/google-project-maven-pentagon.html

Simmel, Georg: *Suurkaupunki ja moderni elämä: Kirjoituksia vuosilta 1895–1917*. Translated into Finnish by Tiina Huuhtanen (Gaudeamus: 2005).

Singal, Jesse: "Social Media Is Making Us Dumber. Here's Exhibit A." *The New York Times*, 11 Jan 2018. Available at: https://www.nytimes.com/2018/01/11/opinion/social-media-dumber-steven-pinker.html

Singer, Natasha: "The Next Privacy Battle in Europe Is Over This New Law," *The New York Times, 27 May 2018. Available at:* https://www.nytimes.com/2018/05/27/technology/europe-eprivacy-regulation-battle.html

Singer, Peter W.: "Robots at War: The New Battlefield," *The Wilson Quarterly*, Winter 2009. Available at: https://wilsonquarterly.com/quarterly/winter-2009-robots-at-war/robots-at-war-the-new-battlefield/

Slamecka, Norman J. & Graf, Peter: The Generation Effect: Delineation of a Phenomenon. *Journal of Experimental Psychology: Human Learning and Memory*, November 1978, 592–604. Available at: https://www.researchgate.net/publication/232485723_The_Generation_Effect_Delineation_of_a_Phenomenon

Smith, Aaron: "US Smartphone Use in 2015," Pew Research Center, 1 Apr 2015. Available at: http://www.pewinternet.org/2015/04/01/us-smartphone-use-in-2015/

Smith, Aaron & Anderson, Janna: "AI, Robotics, and the Future of Jobs," Pew Research Center, 6 Aug 2014. Available at:

http://www.pewinternet.org/2014/08/06/future-of-jobs/

Smith, Noah: "We'll get Flying Cars to Go With Our 140 Characters," Bloomberg, 26 Sep 2017. Available at: https://www.bloomberg.com/view/articles/2017-09-26/we-ll-get-flying-cars-to-go-with-our-140-characters

Solon, Olivia: "The rise of robots: forget evil AI – the real risk is far more insidious," *The Guardian* 30 Aug 2016. Available at: https://www.theguardian.com/technology/2016/aug/30/rise-of-robots-evil-artificial-intelligence-uc-berkeley

Solon, Olivia: "Ex-Facebook president Sean Parker: site made to exploit human 'vulnerability,'" *The Guardian*, 9 Nov 2017. Available at: https://www.theguardian.com/technology/2017/nov/09/facebook-sean-parker-vulnerability-brain-psychology

Sommeiller, Estelle & Price, Mark: *The New Gilded Age: Income Inequality in the US by State, Metropolitan Area, and County* (Economic Policy Institute: 2018).

Sparrow, B.; Liu, J. & Wegner, D. M.: Google effects on memory: cognitive consequences of having information at our fingertips. *Science*, 5. August, 2011, 776–778. Available at: https://www.ncbi.nlm.nih.gov/pubmed/21764755/

Specia, Megan & Mozur, Paul: "A War of Words Puts Facebook at the Center of Myanmar's Rohingya Crisis," *The New York Times*, 27 Oct 2017. Available at: https://www.nytimes.com/2017/10/27/world/asia/myanmar-government-facebook-rohingya.html?mc=adintl&mcid=keywee&mccr=EU&ad-keywords=IntlAudDev&kwp_0=565435&kwp_4=2037985&kwp_1=851001&mtrref=www.facebook.com&module=ArrowsNav&contentCollection=Asia%20Pacific&action=keypress&region=FixedLeft&pgtype=article

Statista: "Number of Hilton Worldwide employees from 2013 to 2017 (in thousands)," 2018. Available at: https://www.statista.com/statistics/297758/number-of-hilton-worldwide-employees/

Statistics Finland: *Tuloerojen kehitys Suomessa 1966-2016* (2017). Available at: https://www.stat.fi/til/tjkt/2016/01/tjkt_2016_01_2017-12-20_kat_001_fi.html

Stephens-Davidowitz, Seth: *Everybody Lies: What the Internet Can Tell Us About Who We Really Are* (Bloomsbury: 2017).

Stevenson, Betsey & Wolfers, Justin: Economic Growth and Subjective Well-Being: Reassessing the Easterlin Paradox, Brookings Papers on Economic Activity, Economic Studies Program, Brookings Institution, Spring 2008, 1–102. Available at: http://www.nber.org/papers/w14282.pdf

Stop Killer Robots Campaign: "Artificial intelligence experts call for ban," 2015. Available at: https://www.stopkillerrobots.org/2015/07/aicall/

Stothart, C.; Mitchum, A & Yehnert, C.: The attentional cost of receiving a cell phone notification. *Journal of Experimental Psychology: Human Perception and Performance,* elokuu 2015, 893–897. Available at: https://www.researchgate.net/publication/279457726_The_Attentional_Cost_of_Receiving_a_Cell_Phone_Notification

Streitfeld, David; Singer, Natasha & Erlanger, Steven: "How Calls for Privacy May Upend Business for Facebook and Google," *The New York Times*, 24 Mar 2018. Available at: https://www.nytimes.com/2018/03/24/technology/google-facebook-data-privacy.html

Subramanian, Samanth: "Inside the Macedonian fake-news complex," *Wired*, 15 Feb 2017. Available at: https://www.wired.com/2017/02/veles-macedonia-fake-news/

Summers, Lawrence H.: "The robots are coming, whether Trump's Treasury secretary admits it or not," the *Washington Post*, 27 Mar 2018. Available at: https://www.washingtonpost.com/news/wonk/wp/2017/03/27/larry-summers-mnuchins-take-on-artificial-intelligence-is-not-defensible/?utm_term=.1322b1e949e3

Sunstein, Cass R.: *#republic: Divided Democracy in the Age of Social*

Media (Princeton University Press: 2017).
Talsi, Noora: *Kodin koneet: Teknologioiden kotouttaminen, käyttö ja vastustus* (Dissertation, University of Eastern Finland: 2014).
The National Non-Discrimination and Equality Tribunal: "Yhdenvertaisuus- ja tasa-arvolautakunta kielsi luottoyhtiötä käyttämästä tilastollista menetelmää luotonhakijaa syrjivästi kuluttajaluotonannossa," 25 Apr 2018. Available at: http://yvtltk.fi/fi/index/tiedotteet/2018/04/yhdenvertaisuus-jatasa-arvolautakuntakielsiluottoyhtiotakayttamastatilastollistamenetelmaaluotonhakijaasyrjivastikuluttajaluotonannossa.html
The Physics ArXiv Blog: "The Face Recognition Algorithm That Finally Outperforms Humans," Medium, 23 Apr 2014. Available at: https://medium.com/the-physics-arxiv-blog/the-face-recognition-algorithm-that-finally-outperforms-humans-2c567adbf7fc
Thiel, Peter: "The End of Future," *National Review,* 3 Oct 2011, Available at: http://www.nationalreview.com/article/278758/end-future-peter-thiel;
Thoreau, Henry David: Walden: Elämää metsässä. Original English *Walden; or, Life in the Woods,* 1854. Translated into Finnish by Antti Immonen (Kotimaa/Kirjapaja: 2010).
Thornton, B.; Faires, A.; Robbins, M. & Rollins, E: The mere presence of a cell phone may be distracting: Implications for attention and task performance. *Social Psychology,* November2014, 479–488. Available at: http://psycnet.apa.org/record/2014-52302-001
Tiainen, Pekka: *Työllisyysaikasarjat* (Ministry of Economic Affairs and Employment: 2014).
Tiku, Nitasha: "How Big Tech Become a Bi-Partisan Whipping Boy," *Wired,* 23 Oct 2017. Available at: https://www.wired.com/story/how-big-tech-became-a-bipartisan-whipping-boy/
Timberg, Craig; Romm, Tony & Dwoskin, Elizabeth: "Facebook:

'Malicious actors' used its tools to discover identities and collect data on a massive global scale," *The Washington Post*, 4 Apr 2018. Available at: https://www.washingtonpost.com/news/the-switch/wp/2018/04/04/facebook-said-the-personal-data-of-most-its-2-billion-users-has-been-collected-and-shared-with-outsiders/?utm_term=.01a8497c5e61

Ting, Deanne: "Airbnb's Latest Investment Values It as much as Hilton and Hyatt Combined," Skift, 23 Sep 2016. Available at: https://skift.com/2016/09/23/airbnbs-latest-investment-values-it-as-much-as-hilton-and-hyatt-combined/

Toffler, Alvin: *Future Shock* (Pan Books: 1972; orig. 1970).

Toffler, Alvin: *The Third Wave* (Bantam Books: 1981).

Trefis Team: "Breaking Down Uber's Valuation: An Interactive Analysis," *Forbes*, 22 Feb 2018. Available at: https://www.forbes.com/sites/greatspeculations/2018/02/22/breaking-down-ubers-valuation-an-interactive-analysis/#69b2c78b4785

Tromholt, M.: The Facebook Experiment: Quitting Facebook Leads to Higher Levels of Well-Being. *Cyberpsychology, Behavior, and Social Networking*, November 2016, 661–666. Available at: https://www.ncbi.nlm.nih.gov/pubmed/27831756

Tufekci, Zeynep: "Beware the Smart Campaign," *The New York Times*, 16 Nov 2012. Available at: https://www.nytimes.com/2012/11/17/opinion/beware-the-big-data-campaign.html

Tufekci, Zeynep: "Facebook's Surveillance Machine," *The New York Times*, 19 Mar 2018. Available at: https://www.nytimes.com/2018/03/19/opinion/facebook-cambridge-analytica.html?rref=collection%2Fcolumn%2FZeynep%20Tufekci&action=click&contentCollection=Opinion&module=Collection&region=Marginalia&src=me&version=column&pgtype=article

Turkle, Sherry: *Simulation and its Discontents* (The MIT Press: 2009).

Turkle, Sherry: *Alone Together: Why We Expect More from Technology and Less From Each Other* (Basic Books: 2012, orig. 2011).

Turkle, Sherry: *Reclaiming Conversation: The Power of Talk in a Digital Age* (Penguin Press: 2015).

Twenge, Jean M.: "Have Smartphones Destroyed a Generation," *The Atlantic,* September 2017. Available at: https://www.theatlantic.com/magazine/archive/2017/09/has-the-smartphone-destroyed-a-generation/534198/

Uber website, 2018. Available at: https://www.uber.com/newsroom/company-info/

Uhls, Yalda T.; Michikyan, Minas; Morris, Jordan; Garcia, Debra; Small, Gary W.; Zgourou, Eleni & Greenfield, Patricia M.: Five days at outdoor education camp without screens improves preteen skills with nonverbal emotion cues. *Computers in Human Behavior,* October 2014, 387–392. Available at: https://www.sciencedirect.com/science/article/pii/S0747563214003227

Uncapher, M. R.; Thieu, M; Wagner, A. D.: Media multitasking and memory: Differences in working memory and long-term memory. *Psychonomic Bulletin & Review,* April 2016, 483–490. Available at: https://www.ncbi.nlm.nih.gov/pubmed/26223469

US Army: *Robotics and Autonomous Systems Strategy* (2017). Available at: http://www.tradoc.army.mil/FrontPageContent/Docs/RAS_Strategy.pdf

US Department of Defense: *Unmanned Systems Integrated Roadmap, FY2011-2036* (2011). Available at: https://fas.org/irp/program/collect/usroadmap2011.pdf

US Department of Defense: *Directive: Autonomy in Weapon Systems* (2012). Available at: http://www.esd.whs.mil/Portals/54/Documents/DD/issuances/dodd/300009p.pdf

US Government: *Humanitarian benefits of emerging technologies in the area of lethal autonomous weapon systems,* Group of Governmental Experts of the High Contracting Parties to the Convention on Prohibitions or Restrictions on the Use of Certain Conventional Weapons Which May Be Deemed to Be Excessively Injurious or to Have Indiscriminate Effects (2018). Available at: https://www.unog.ch/80256EDD006B8954/(httpAssets)/7C177AE5BC10B588C125825F004B06BE/$file/CCW_GGE.1_2018_WP.4.pdf

Vahvanen, Pekka: "Kirjatutka: Alati jatkuva itä–länsi-ottelu," *Ulkopolitiikka*, 2/2011. Available at: http://www.ulkopolitiikka.fi/artikkeli/849/kirjatutka_alati_jatkuva_it_l_nsi-ottelu/

Vahvanen, Pekka: "Elämässämme on liikaa nautintoja ja liian vähän kykyä nauttia," *Helsingin Sanomat*, 15 Jun 2014. Available at: https://www.hs.fi/sunnuntai/art-2000002739169.html

Vahvanen, Pekka: "Jos alamme tuomita toisiamme kuvottavien mutta yksityisiksi tarkoitettujen sanomisten vuoksi, olemme matkalla takaisin keskiajalle," *Helsingin Sanomat* 23 Jul 2017. Available at: https://www.hs.fi/sunnuntai/art-2000005298654.html

Valinsky, Jordan: "Even the iPhone's designer is worried about phone addiction," CNN, 9 Jan 2018. Available at: https://money.cnn.com/2018/01/09/technology/business/apple-iphone-addiction-tony-fadell/index.htmlVan der Weyde, William M.: *The Life and Works of Thomas Paine* (The University of Chicago Press: 1925), Volume 3, Article 1, Section 9, Clause 8, Document 2. Available at: http://press-pubs.uchicago.edu/founders/documents/a1_9_8s2.html

Vänttinen, Alisa: "Asiantuntija: 'Yksinäisyys on vakavampi ongelma kuin aikaisemmin,'" *Savon Sanomat*, 22 Aug 2016. Available at: https://www.savonsanomat.fi/savo/Asiantuntija-Yksin%C3%A4isyys-on-vakavampi-ongelma-kuin-aikaisemmin/820914

Verduyn, P.; Lee, D. S.; Park, J.; Shablack, H.; Orvell, A.; Bayer, J; Ybarra, O; Jonides, J. & Kross, E.: Passive Facebook usage undermines affective well-being: Experimental and longitudinal evidence. *Journal of Experimental Psychology. General*,

April 2015, 480–488. Available at: https://www.ncbi.nlm.nih.gov/pubmed/25706656

Verduyn, Philippe; Ybarra, Oscar; Résibois, Maxime; Jonides, John & Kross, Ethan: Do Social Network Sites Enhance or Undermine Subjective Well-Being? A Critical Review, *Social Issues and Policy Review*, January 2017, 274–302. Available at: https://spssi.onlinelibrary.wiley.com/doi/full/10.1111/sipr.12033

Vincent, James: "Twitter taught Microsoft's AI chatbot to be a racist asshole in less than a day," The Verge, 24 Mar 2016. Available at: https://www.theverge.com/2016/3/24/11297050/tay-microsoft-chatbot-racist

Vincent, James: "Putin says the nation that leads in AI 'will be the ruler of the world,'" The Verge, 4 Sep 2017a. Available at: https://www.theverge.com/2017/9/4/16251226/russia-ai-putin-rule-the-world

Vincent, James: "DeepMind's Go-playing AI doesn't need human help to beat us anymore," The Verge, 18 Oct 2017b. Available at: https://www.theverge.com/2017/10/18/16495548/deepmind-ai-go-alphago-zero-self-taught

Vincent, James: "Most Americans think artificial intelligence will destroy other people's jobs, not theirs," The Verge, 7 Mar 2018. Available at https://www.theverge.com/2018/3/7/17089904/ai-job-loss-automation-survey-gallup

Vinge, Vernor: *The Coming Technological Singularity: How to Survive in the Post-Human Era*. Vision-21 Symposium, 30-31 Mar 1993. Available at: https://edoras.sdsu.edu/~vinge/misc/singularity.html

Vinter, Phil: "Zadie Smith pays tribute to computer software that BLOCKS internet sites allowing her to write new book without distractions," *Daily Mail*, 1 Sep 2012. Available at: http://www.dailymail.co.uk/news/article-2196718/Zadie-Smith-pays-tribute-software-BLOCKS-internet-sites-allowing-write-new-book-distractions.html

Vlahos, James: "A Son's Race to Give his Dying Father Artificial Immortality," *Wired*, 18 Jul 2017. Available at: https://www.wired.com/story/a-sons-race-to-give-his-dying-father-artificial-immortality/

Wachter, Sandra; Mittelstadt, Brent; Floridi, Luciano: Why a Right to Explanation of Automated Decision-Making Does Not Exist in the General Data Protection Regulation. *International Data Privacy Law*. 28 Dec 2016. Available at: https://papers.ssrn.com/sol3/papers.cfm?abstract_id=2903469

Walsh, Bryan: "The Surprisingly Large Energy Footprint of the Digital Economy," *Time*, 14 Aug 2013. Available at: http://science.time.com/2013/08/14/power-drain-the-digital-cloud-is-using-more-energy-than-you-think/

Walters, Guy, "Are we mad to have let a maverick scientist create a virus that could wipe out 400 million people?" *Daily Mail*, 3 Jul 2014. Available at: http://www.dailymail.co.uk/sciencetech/article-2678732/Are-mad-let-maverick-scientist-create-virus-wipe-400-million-people.html

Ward, Mark: "It is easy to expose users' secret web habits, say researchers," BBC, 31 Jul 2017. Available at: https://www.bbc.co.uk/news/technology-40770393

Warrick, Joby: "Use of weaponized drones by ISIS spurs terrorism fears," *The Washington Post*, 21 Feb 2017. Available at: https://www.washingtonpost.com/world/national-security/use-of-weaponized-drones-by-isis-spurs-terrorism-fears/2017/02/21/9d83d51e-f382-11e6-8d72-263470bf0401_story.html?utm_term=.826fd0a70389

Weaver, Matthew: "Facebook scandal: I am being used as scapegoat – academic who mined data," *The Guardian*, 21 Mar 2018. Available at: https://www.theguardian.com/uk-news/2018/mar/21/facebook-row-i-am-being-used-as-scapegoat-says-academic-aleksandr-kogan-cambridge-analytica

Weber, Max: *Protestanttinen etiikka ja kapitalismin henki*. Original German *Die protestantische Ethik und der Geist des Kapitalis-*

mus, 1905. Translated into Finnish by Timo Kyntäjä (WSOY: 1990).

Weisberg, Jacob: "We are hopelessly hooked," *The New York Review of Books*, 25 Feb 2016. Available at: http://www.nybooks.com/articles/2016/02/25/we-are-hopelessly-hooked/

Weizenbaum, Joseph: *Computer Power and Human Reason: From Judgement to Calculation* (Pelican Books: 1984, orig. 1976).

Weller, Chris: "Bill Gates and Steve Jobs raised their kids tech-free – And it should've been a red flag," *The Independent*, 24 Oct 2017. Available at: http://www.independent.co.uk/life-style/gadgets-and-tech/bill-gates-and-steve-jobs-raised-their-kids-tech-free-and-it-shouldve-been-a-red-flag-a8017136.html

Wikiquote: Cardinal Richelieu. Available at: https://en.wikiquote.org/wiki/Cardinal_Richelieu

Wilkinson, Richard & Pickett, Kate: *Tasa-arvo ja hyvinvointi: Miksi tasa-arvo on hyväksi kaikille?* (HS Kirjat 2011; orig. 2009).

Wilmer, Henry H. & Chein, Jason M.: Mobile technology habits: patterns of association among device usage, intertemporal preference, impulse control, and reward sensitivity. *Psychonomic Bulletin and Review*, October 2016, 1607–1614. Available at: https://www.ncbi.nlm.nih.gov/pubmed/26980462

Wilmer, Henry H.; Sherman, Lauren E. & Chein, Jason M.: Smartphones and Cognition: A Review of Research Exploring the Links between Mobile Technology Habits and Cognitive Functioning. *Frontiers in Psychology*, April 2017, 1–16. Available at: https://www.ncbi.nlm.nih.gov/pmc/articles/PMC5403814/#B122 s

Witte, Griff: "Snowden says government spying worse than Orwellian," *The Washington Post*, 25 Dec 2013. Available at: https://www.washingtonpost.com/world/europe/snowden-says-spying-worse-than-orwellian/2013/12/25/e9c806aa-6d90-11e3-a5d0-6f31cd74f760_story.html?utm_term=.e6a1eed5002d

Wolf, M. & Barzillai, M.: The importance of deep reading. *Educational Leadership*, March 2009, 32–37.

Wong, Julia Carrie: "Facebook blocks Chechnya activist page in latest case of wrongful censorship," *The Guardian*, 6 Jun 2017. Available at: https://www.theguardian.com/technology/2017/jun/06/facebook-chechnya-political-activist-page-deleted

Wong, Julia Carrie: "Former Facebook executive: social media is ripping society apart," *The Guardian*, 12 Dec 2017. Available at: https://www.theguardian.com/technology/2017/dec/11/facebook-former-executive-ripping-society-apart

Wright, Georg Henrik von: Progress: Fact and Fiction, Burgen, Arnold; McLaughlin, Peter & Mittelstraß, Jürgen (ed.): *The Idea of Progress* (de Gruyter: 1997), 19–46.

Wright, Ronald: *A Short History of Progress*, (House of Anansi Press: 2004).

Wu, Xiaolin & Zhang, Xi: Automated Inference on Criminality using Face Images. *arXiv*, 21 Nov 2016. Available at: https://arxiv.org/pdf/1611.04135v2.pdf

Wu, Youyou; Kosinski, Michal & Stillwell, David: Computer-based personality judgments are more accurate than those made by humans. *Proceedings of the National Academy of Sciences of the United States of America*, 27 Jan 2015, 1036–1040. Available at: http://www.pnas.org/content/112/4/1036

Yates, Eames: "This Silicon Valley school shuns technology — yet most of the students are children of tech execs," *Business Insider*, 23 Mar 2017: Available at: http://www.businessinsider.com/waldorf-silicon-valley-school-shuns-technology-2017-3?r=US&IR=T&IR=T

Yoon, Sung-won: "Korea takes first step to introduce 'robot tax,'" *The Korea Times*, 7 Aug 2017. Available at: http://www.koreatimes.co.kr/www/news/tech/2017/08/133_234312.html

Young, Cristobal: Losing a Job: The Nonpecuniary Cost of Unemployment in the United States. *Social Forces*, November

2012, 609–633. Available at: https://sociology.stanford.edu/sites/default/files/publications/losing_a_job-the_non-pecuniary_cost_of_unemployment_in_the_united_states.pdf

Yudkowsky, Eliezer: *Artificial Intelligence as a Positive and Negative Factor in Global Risk* (Machine Intelligence Research Institute: 2008). Available at: https://intelligence.org/files/AIPosNegFactor.pdf

Zittrain, Jonathan: "Facebook could Decide an Election Without Anyone Ever Finding Out," *The New Republic*, 2 Jun 2014. Available at: https://newrepublic.com/article/117878/information-fiduciary-solution-facebook-digital-gerrymandering

zer0
books

CULTURE, SOCIETY & POLITICS

The modern world is at an impasse. Disasters scroll across our smartphone screens and we're invited to like, follow or upvote, but critical thinking is harder and harder to find. Rather than connecting us in common struggle and debate, the internet has sped up and deepened a long-standing process of alienation and atomization. Zer0 Books wants to work against this trend.
With critical theory as our jumping off point, we aim to publish books that make our readers uncomfortable. We want to move beyond received opinions.
Zer0 Books is on the left and wants to reinvent the left. We are sick of the injustice, the suffering and the stupidity that defines both our political and cultural world, and we aim to find a new foundation for a new struggle.

If this book has helped you to clarify an idea, solve a problem or extend your knowledge, you may want to check out our online content as well. Look for Zer0 Books: Advancing Conversations in the iTunes directory and for our Zer0 Books YouTube channel.

Popular videos include:

Žižek and the Double Blackmain

The Intellectual Dark Web is a Bad Sign

Can there be an Anti-SJW Left?

Answering Jordan Peterson on Marxism

Follow us on Facebook
at https://www.facebook.com/ZeroBooks and Twitter at https://
twitter.com/Zer0Books

Bestsellers from Zer0 Books include:

Give Them An Argument
Logic for the Left
Ben Burgis
Many serious leftists have learned to distrust talk of logic. This is a serious mistake.
Paperback: 978-1-78904-210-8 ebook: 978-1-78904-211-5

Poor but Sexy
Culture Clashes in Europe East and West
Agata Pyzik
How the East stayed East and the West stayed West.
Paperback: 978-1-78099-394-2 ebook: 978-1-78099-395-9

An Anthropology of Nothing in Particular
Martin Demant Frederiksen
A journey into the social lives of meaninglessness.
Paperback: 978-1-78535-699-5 ebook: 978-1-78535-700-8

In the Dust of This Planet
Horror of Philosophy vol. 1
Eugene Thacker
In the first of a series of three books on the Horror of Philosophy, *In the Dust of This Planet* offers the genre of horror as a way of thinking about the unthinkable.
Paperback: 978-1-84694-676-9 ebook: 978-1-78099-010-1

The End of Oulipo?
An Attempt to Exhaust a Movement
Lauren Elkin, Veronica Esposito
Paperback: 978-1-78099-655-4 ebook: 978-1-78099-656-1

Capitalist Realism
Is There No Alternative?
Mark Fisher
An analysis of the ways in which capitalism has presented itself as the only realistic political-economic system.
Paperback: 978-1-84694-317-1 ebook: 978-1-78099-734-6

Rebel Rebel
Chris O'Leary
David Bowie: every single song. Everything you want to know, everything you didn't know.
Paperback: 978-1-78099-244-0 ebook: 978-1-78099-713-1

Kill All Normies
Angela Nagle
Online culture wars from 4chan and Tumblr to Trump.
Paperback: 978-1- 78535-543-1 ebook: 978-1-78535-544-8

Cartographies of the Absolute
Alberto Toscano, Jeff Kinkle
An aesthetics of the economy for the twenty-first century.
Paperback: 978-1-78099-275-4 ebook: 978-1-78279-973-3

Malign Velocities
Accelerationism and Capitalism
Benjamin Noys
Long listed for the Bread and Roses Prize 2015, *Malign Velocities* argues against the need for speed, tracking acceleration as the symptom of the ongoing crises of capitalism.
Paperback: 978-1-78279-300-7 ebook: 978-1-78279-299-4

Meat Market
Female Flesh under Capitalism
Laurie Penny
A feminist dissection of women's bodies as the fleshy fulcrum of capitalist cannibalism, whereby women are both consumers and consumed.
Paperback: 978-1-84694-521-2 ebook: 978-1-84694-782-7

Babbling Corpse
Vaporwave and the Commodification of Ghosts
Grafton Tanner
Paperback: 978-1-78279-759-3 ebook: 978-1-78279-760-9

New Work New Culture
Work we want and a culture that strengthens us
Frithjoff Bergmann
A serious alternative for mankind and the planet.
Paperback: 978-1-78904-064-7 ebook: 978-1-78904-065-4

Romeo and Juliet in Palestine
Teaching Under Occupation
Tom Sperlinger
Life in the West Bank, the nature of pedagogy and the role of a university under occupation.
Paperback: 978-1-78279-637-4 ebook: 978-1-78279-636-7

Ghosts of My Life
Writings on Depression, Hauntology and Lost Futures
Mark Fisher
Paperback: 978-1-78099-226-6 ebook: 978-1-78279-624-4

Sweetening the Pill
or How We Got Hooked on Hormonal Birth Control
Holly Grigg-Spall
Has contraception liberated or oppressed women?
Sweetening the Pill breaks the silence on the dark side of hormonal contraception.
Paperback: 978-1-78099-607-3 ebook: 978-1-78099-608-0

Why Are We The Good Guys?
Reclaiming Your Mind from the Delusions of Propaganda
David Cromwell
A provocative challenge to the standard ideology that Western power is a benevolent force in the world.
Paperback: 978-1-78099-365-2 ebook: 978-1-78099-366-9

The Writing on the Wall
On the Decomposition of Capitalism and its Critics
Anselm Jappe, Alastair Hemmens
A new approach to the meaning of social emancipation.
Paperback: 978-1-78535-581-3 ebook: 978-1-78535-582-0

Enjoying It
Candy Crush and Capitalism
Alfie Bown
A study of enjoyment and of the enjoyment of studying. Bown asks what enjoyment says about us and what we say about enjoyment, and why.
Paperback: 978-1-78535-155-6 ebook: 978-1-78535-156-3

Color, Facture, Art and Design
Iona Singh
This materialist definition of fine-art develops guidelines for architecture, design, cultural-studies and ultimately social change.
Paperback: 978-1-78099-629-5 ebook: 978-1-78099-630-1

Neglected or Misunderstood
The Radical Feminism of Shulamith Firestone
Victoria Margree
An interrogation of issues surrounding gender, biology, sexuality, work and technology, and the ways in which our imaginations continue to be in thrall to ideologies of maternity and the nuclear family.
Paperback: 978-1-78535-539-4 ebook: 978-1-78535-540-0

How to Dismantle the NHS in 10 Easy Steps (Second Edition)
Youssef El-Gingihy
The story of how your NHS was sold off and why you will have to buy private health insurance soon. A new expanded second edition with chapters on junior doctors' strikes and government blueprints for US-style healthcare.
Paperback: 978-1-78904-178-1 ebook: 978-1-78904-179-8

Digesting Recipes
The Art of Culinary Notation
Susannah Worth
A recipe is an instruction, the imperative tone of the expert, but this constraint can offer its own kind of potential. A recipe need not be a domestic trap but might instead offer escape – something to fantasise about or aspire to.
Paperback: 978-1-78279-860-6 ebook: 978-1-78279-859-0